Chemical Engineering Methods and Technology

nova
science publishers
www.novapublishers.com

Chemical Engineering Methods and Technology

Chemical Vapor Deposition (CVD): Methods and Technologies
Levi Karlsson (Editor)
2021. ISBN: 978-1-53619-949-9 (Hardcover)
2021. ISBN: 978-1-53619-990-1 (eBook)

Fluidized Bed Reactors: Principles and Applications
Kristian Beike (Editor)
2020. ISBN: 978-1-53617-727-5 (Softcover)
2020. ISBN: 978-1-53617-733-6 (eBook)

Microfiltration: Principles, Process and Applications
Youssef El Rayess, PhD (Editor)
2019. ISBN: 978-1-53616-320-9 (Hardcover)
2019. ISBN: 978-1-53616-347-6 (eBook)

Zirconia and Alumina Bioceramic Biocompatibility
Giulio Maccauro, Pierfrancesco Rossi Iommetti,
Luca Raffaelli and Paolo Francesco Manicone (Editors)
2012. ISBN: 978-1-62081-650-9 (Online Book)

Perchlorates: Production, Uses and Health Effects
Lawrence E. Matthews (Editor)
2011. ISBN: 978-1-61761-812-3 (Hardcover)
2011. ISBN: 978 (eBook)

More information about this series can be found at
https://novapublishers.com/product-category/series/chemical-engineering-methods-and-technology/

Amanda G. Carlton
Editor

Properties and Uses of Calcium Silicate

nova
science publishers
www.novapublishers.com

Library of Congress Cataloging-in-Publication Data

ISBN: 979-8-88697-128-6

Published by Nova Science Publishers, Inc. † New York

Contents

Preface

Calcium silicate is a compound that can be derived from limestone and diatomaceous earth and has applications as an anticaking agent in food preparation, as insulation, and more. The four chapters of this book explore the various characteristics and applications of calcium silicate. Chapter One examines the effectiveness of calcium-silicate cements in furcal perforation repair. Chapter Two investigates methods of producing cement with lower energy consumption by using calcium silicate. Chapter Three deals with the use of synthesized calcium silicates in compositions for the restoration of buildings. Lastly, Chapter Four evaluates and compares the effects of simulated tissue fluid on the Berkocivh microhardness of four calcium silicate cements.

Chapter 1 - Calcium-silicate cements (CSCs) are the most commonly used root perforation repair materials. One of the major factors determining the healing process is the hermeticity of sealing provided by the material. The aim of this study was to evaluate the furcal sealing ability and marginal adaptation of five CSCs and the possible correlation between the two assessment methods.

A total of 50 molars with well-defined furcations were randomly divided into five equal groups of 10 teeth each. Artificial perforations were performed in the pulpal floor center. The perforations were sealed with one of the five CSCs as follows: white MTA-Angelus, white ProRoot MTA, Aureoseal, BioAggragate, and Biodentine. After coronal restoration of endodontic cavities the teeth were immersed in 0.2% Rhodamine B dye for 72 hours. They were afterwards rinsed and sectioned longitudinally. The depth of linear dye penetration was measured to the nearest 0.0x mm under a stereomicroscope. Subsequently, the extent of marginal adaptation was evaluated and measured by SEM.

No dye penetration was registered in ProRoot MTA, Aureoseal and BioAggragate. MTA-Angelus showed greater dye penetration compared to other CSCs. The marginal adaptation of all materials was very similar, and the produced gap size ranged from 5.62-7.28 μm without statistically significant differences between the groups. A direct correlation was found between the two assessment methods of the hermeticity of sealing.

Conclusion: Under the conditions of this study, all calcium silicate-based cements, except wMTA-Angelus, showed good marginal adaptation with absence or minimal dye microleakage and are proven to be suitable for furcal perforation repair.

Chapter 2 - The term "low-energy cements" describes cements that can replace ordinary Portland cement in at least some applications and can also reduce energy consumption during the production process. Extension of these cements could lead both to lower costs of binders' production and lower emissions, CO2 especially. Belite cements are part of low-energy cements. Pure belite clinkers with interstitial matter comprising C3A and C4AF are not produced, since they yield materials with insufficient strength. Currently, sulfoaluminate-belite cements, low-energy cements doped with fluorine and chlorine as well as undoped high-belite cements are produced in limited volumes in China. Results of hydraulic activation of high-belite clinker by sulfate anions are also given in this paper. The principle of activation is the preparation of belite clinker with increased Ca:Si ratio in the structure of dicalcium silicate by substitution a part of SiO44- by SO42-. Cements prepared from these high-belite clinkers with alite content up to 30% that were burned at 1,350°C display the same technological properties as ordinary alite Portland cements, including early strengths as ordinary alite Portland cements and comply with standard EN 197-1.

Chapter 3 - Lime mixes are the primary material for restoration work in historic buildings. The research object of this study is modifying lime binders with specially synthesized calcium silicate hydrates (CSHs). This study aimed to improve lime render mixes' weather resistance. Hydro silicates of calcium (HSC) are received by synthesis from liquid sodium glass in the presence of calcium chloride CaCl2 with the subsequent drying of sediment at a temperature 105-300°C and its crushing. Studying of material composition of the filler, made by methods of the X-ray phase analysis (RFA) and infrared (IK) spectroscopy, and also structure on the electronic scanning microscope of PhenomTM G2 pro showed the existence in synthesizable material the calcium silicates of various basicity.

The conducted research show that the filler based on HSC possesses hydraulic activity. The activity of filler, defined according to a technique [5], makes 195-250 mg/g depending on the synthesis mode.

The following factors were considered: the density of the liquid glass, the amount of the precipitating additive, the rate of introduction of the precipitating additive, the drying mode of the precipitate, and the storage time of the precipitate. The research methods were X-ray diffraction analysis, differential thermal analysis, Fourier transform infrared spectrometry, and optical and electron microscopy. Regularities of activity change of filler interaction with lime depending on conditions of its receiving are revealed. It was revealed that lime compositions with CSH have a higher strength gain rate than the control compositions. Decrease of amount of free lime in limy samples with additive of synthesizable hydro silicates, increase of durability are revealed. A mathematical model of the kinetics of hardening a lime composite based on a binder filled with CSH was obtained. The regularities of the change in the lime composite's strength depending on the filler grinding's fineness, its content, and the amount of mixing water have been established. It was revealed that the introduction of CSH into the lime composition increases the weather resistance of facade lime mixtures by reducing the porosity and increasing the volume of closed pores of the composite.

Chapter 4 - Calcium-silicate cements (CSCs) are bioactive materials with a variety of applications in dentistry such as direct pulp capping, perforation repair, retrograde root-end filling, regenerative therapy, etc. Under clinical condition, these materials come into contact with tissue fluids and blood.

The aim of the present study was to evaluate and compare the effects of simulated tissue fluid (STF) on the Berkovich microhardness (HB) of four CSCs - white ProRoot MTA, gray MTA-Angelus, BioAggragate, and Biodentine. After setting, the samples were immersed in STF for different periods of time. Changes were assessed over a period of 24 hours and 30 days. Twelve measurements were made on each sample. A total of 80 samples were examined.

At the 24th hour Biodentine showed the greatest mean microhardness value, but a significant reduction in its microhardness was found from 0.74 GPa (±0.07) to 0.48 GPa (±0.07) after 30 days of exposure to STF. A significant increase of microhardness of MTA-Angelus and Bioaggregate was found at 30th day.

Conclusion: The contact with STF for a period of 30 days led to significant changes of the microhardness of all tested materials, except for ProRoot MTA.

Chapter 1

Comparative Evaluation of the Hermeticity of Sealing of Calcium-Silicate Cements Used in Furcal Perforation Repair

I. Dimitrova*, PhD
and Y. Kouzmanova, PhD
Department of Conservative Dentistry,
Medical University, Sofia, Bulgaria

Abstract

Calcium-silicate cements (CSCs) are the most commonly used root perforation repair materials. One of the major factors determining the healing process is the hermeticity of sealing provided by the material. The aim of this study was to evaluate the furcal sealing ability and marginal adaptation of five CSCs and the possible correlation between the two assessment methods.

A total of 50 molars with well-defined furcations were randomly divided into five equal groups of 10 teeth each. Artificial perforations were performed in the pulpal floor center. The perforations were sealed with one of the five CSCs as follows: white MTA-Angelus, white ProRoot MTA, Aureoseal, BioAggragate, and Biodentine. After coronal restoration of endodontic cavities the teeth were immersed in 0.2% Rhodamine B dye for 72 hours. They were afterwards rinsed and sectioned longitudinally. The depth of linear dye penetration was measured to the nearest 0.0x mm under a stereomicroscope. Subsequently, the extent of marginal adaptation was evaluated and measured by SEM.

* Corresponding Author's Email: vanja_ves@abv.bg.

In: Properties and Uses of Calcium Silicate
Editor: Amanda G. Carlton
ISBN: 979-8-88697-128-6

No dye penetration was registered in ProRoot MTA, Aureoseal and BioAggragate. MTA-Angelus showed greater dye penetration compared to other CSCs. The marginal adaptation of all materials was very similar, and the produced gap size ranged from 5.62-7.28 μm without statistically significant differences between the groups. A direct correlation was found between the two assessment methods of the hermeticity of sealing.

Conclusion: Under the conditions of this study, all calcium silicate-based cements, except wMTA-Angelus, showed good marginal adaptation with absence or minimal dye microleakage and are proven to be suitable for furcal perforation repair.

Keywords: calcium-silicate cements, sealing ability, marginal adaptation, furcal perforation

Introduction

Twenty-first-century modern endodontics has evolved significantly in many aspects, and these recent advances have greatly improved the healing outcome and success rate of both conservative and surgical endodontic treatment (Camilleri, 2020). Over the last three decades, great interest and activity have been shown in the development of bioactive dental materials with the ability to interact and promote regeneration of vital tissues (Zafar, Jamal, and Ghafoor, 2020). The first such material was Mineral trioxide aggregate (MTA). It was introduced by Torabinejad et al., in 1993 and was intended specifically for endodontic purposes (Lee, Monsef, and Torabinejad, 1993; Torabinejad et al., 1993).

MTA was a revolutionary material that opened the door to a new era for hydraulic cements and regenerative endodontics (Camilleri et al., 2022; Kahler, Chugal, and Lin, 2017). The advent of this brand had a huge impact on the development of biomaterials (Ha et al., 2017; Zafar, Jamal, and Ghafoor, 2020). Nowadays, it is regarded as a gold standard repair material for a wide range of clinical applications, including pulp capping, pulpotomy, apexification, perforation repair, retrograde filling, resorption repair, etc. (Camilleri et al., 2022; Cervino et al., 2020; Kahler, Chugal, and Lin, 2017; Parirokh and Torabinejad, 2018; Torabinejad, Parirokh, and Dummer, 2018).

Numerous experimental and clinical investigations have been performed on it since its introduction until now (Parirokh and Torabinejad, 2010a; Parirokh and Torabinejad, 2010b; Parirokh and Torabinejad, 2018; Torabinejad and Parirokh, 2010; Torabinejad, Parirokh, and Dummer, 2018).

Generally, the obtained results indicate that MTA may be considered to be an almost ideal endodontic material due to its unique properties. It hardens in humid conditions, seals the endodontic space almost hermetically, and exhibits minimal microleakage. It is practically insoluble in water and shows minimal expansion when set (Parirokh and Torabinejad, 2010a; Torabinejad and Parirokh, 2010).

Nowadays, several modifications of this material have been made and many newer "bioactive endodontic cements" have been developed. Based on the chemical compositions, they are also referred to as Calcium silicate-based cements (CSCs) (Parirokh and Torabinejad, 2018; Zafar, Jamal, and Ghafoor, 2020). From a clinical point of view, these biomaterials have many advantages. They have shown excellent biocompatibility and bioactivity with osteoinductive, cementoconductive, and cementoinductive properties as well as antibacterial activity (Abboud et al., 2021; Bakhtiar et al., 2017; Nabavizadeh et al., 2022; Parirokh and Torabinejad, 2018; Mahmood Talabani, Taha Garib, and Masaeli, 2020).

Analytical Framework

Literature Review

Endodontic root perforations are complications in the course of endodontic treatment, representing mechanical or pathological communications between the root canal system and the outer root surface (Mitthra et al., 2020; Sarao, Berlin-Broner, and Levin, 2021). Iatrogenic perforations may occur accidentally at any stage of endodontic treatment. They are a result of procedural errors in the preparation of the endodontic access, the search for orifices, improper root canal shaping, or insertion of intra-radicular posts (Alamoudi et al., 2019; Goldberg, 2020; Kouzmanova, 2019).

The successful perforation repair is crucial because, if not treated, perforation can lead to treatment failure and even tooth loss (Bakhtiar et al., 2017, Goldberg, 2020, Sarao, Berlin-Broner, and Levin, 2021). Such complications are most commonly observed in molars compared to anterior and premolar teeth (Akbar, 2015; Alamoudi et al., 2019). In addition, Haji-Hassani et al., have found that 22.4% of the total mistakes in root canal treatment procedures are in the lower molars (Haji-Hassani, Bakhshi, and Shahabi, 2015). Therefore, the main goals of perforation treatment are the hermetic sealing of the defect and communication between the endodontium and periodontium, overcoming the inflammatory process and loss of tissue attachment, and complete recovery of the periodontal complex (Birage et al., 2019; Saed, Ashley, and Darsey, 2016).

Historically, before the introduction of MTA, a wide range of materials were applied for endodontic perforation treatment: indium foil, dental amalgam, calcium sulphate, super EBA, IRM, calcium hydroxide, hydroxyapatite-based materials, etc. More recently, newer groups of materials have been preferred, such as glass-ionomer cements (conventional, resin-modified, and metal-modified GIC) and composites (Birage et al., 2019; Goldberg, 2020; Kakani et al., 2015). However, over the last two decades, MTA has been suggested as a material of choice to restore almost all types of perforations (Hegde, Varghese, and Malhotra, 2017; Jayakumar, 2018; Mitthra et al., 2020).

Ideally, the material for perforation repair should be impermeable to moisture, antibacterial, noncorrosive, nontoxic, nonresorbable, easy to manipulate, unaffected by blood, adhesive to dentin, easily adaptable, radiopaque, and cost-effective. It is placed in direct contact with living tissues, and for this reason, it should be biocompatible and should promote the regeneration of the periodontal tissues (Nabavizadeh et al., 2022; Solanki, Venkappa, and Shah, 2018; von Loetzen and Hulsmann, 2018).

The prognosis and healing outcome of root perforations depend on many factors such as their localization and size, the presence of bacterial contamination, proximity to the crestal bone and epithelial attachment, time elapsed, clinician's skills, etc. The choice of suitable repair material has also shown astrong correlation with the healing process (Baroudi and Samir, 2016; Estrela et al., 2018; Hegde, Varghese, and Malhotra, 2017; Mitthra et al., 2020).

Nowadays, the selection of material for this purpose is focused mainly on properties such as hermetic seal ability, stimulating the healing and biological seal by mineralized tissue deposition, and antibacterial potential (fon Loetzen and Hulsmann, 2018; Kahler, Chugal, and Lin, 2017; Mitthra et al., 2020). Due to its hydrophilic nature, MTA andcalcium silicate-based cements can absorb liquids during their setting process and are considered to have the ability to penetrate into dentinal tubules and gaps (Nica et al., 2020). According to some authors, this peculiar property is a result of the material's interaction with dentin and the forming of a mineralized intermediate zone, with tag-like structures that extend into the dentinal tubules (Estrela et al., 2018; Han and Okuji, 2011). These phenomena, along with a slight expansion during hardening and production of crystalline hydroxyapatite in the material-dentin wall interface can provide excellent sealing and marginal adaptation (Ayatollahy et al., 2017; Parirokh and Torabinejad, 2010a; Sarkar et al., 2005).

MTA is the only restorative material which provides not only an excellent and effective physical sealing of root perforations but also a biological seal. Due to its high alkalinity and release of calcium ions, which stimulate differentiation of cells, it has the ability to form newly mineralized tissue (Gandolfi et al., 2017; Hinata et al., 2017; Kunjan and Ballal, 2020; Makhlouf et al., 2020). In addition, it consistently allows cementum overgrowth and may promote periodontal tissue regeneration or repair, in particular of the periodontal ligament and alveolar bone (Aboud et al., 2017; Bakhtiar et al., 2017; Baroudi and Samir, 2016; Mahmood Talabani, Taha Garib, and Masaeli, 2020).

For these reasons, in recent years, MTA and CSCs are considered to be ideal materials for perforation repair. These modern repair materials have significantly improved the healing outcome of root perforations, even those localized in areas with poor prognosis, which in the past were believed to be hopeless (Baroudi and Samir, 2016; von Loetzen and Hulsmann, 2018). Long-term clinical studies have reported good to excellent healing results and a high success rate of perforation recovery in terms of tooth preservation (Aldayri, Nourallah, and Badr, 2019; Krupp et al., 2013; de Melo et al., 2019, Kouzmanova, Dimitrova, and Tsanova-Tosheva, 2021; Mente et al., 2014; Pontius et al., 2013; Salceanu et al., 2017).

As a consequence, the application of older repair materials is significantly reduced because of their inability to seal the perforation site adequately and to prevent communication between the oral cavity and the root canal system of the affected tooth as well as because of their lack of sufficient biocompatibility (Balachandran and Gurucharan, 2013; Birage et al., 2019, Kouzmanova and Nikiforova, 2013, Nandakumar and Nasim, 2018, Torabinejad, Watson, and Pitt Ford, 1993).

CSCs are classified on the basis of their chemical composition, structural analysis, and area of application. According to their specific uses, Camilleri et al. divided the hydraulic cements into three major groups: for intra-coronal, intra-radicular, and extra-radicular applications (Camilleri et al., 2022). Four main generations of these materials are distinguished (Tsanova-Tosheva, Dimitrova, and Kouzmanova, 2021; Zafar, Jamal, and Ghafoor, 2020):

Generation I: Gray ProRoot MTA and white ProRoot MTA;
Generation II: Modifications to MTA (gray and white MTA-Angelus);

Generation III: Calcium phosphate based bioceramics (Bioaggregate, Biodentine, iRootSP - also retailed as EndosequenceBC and SmartPaste Bio, Endo CPM, MTA Obtura, Tech Biosealer Endo, CEM cement, Ortho MTA, Retro MTA, MTA plus, Generex A, Generx B, etc.;

Generation IV: Hybrid cements (resin modified light-cured cements) - TheraCal LC, BioCal-Cap, Calcium phosphate/Calcium silicate cement, NRC, MTA with 4-META/MMA-TBB.

Material Characteristics

ProRoot MTA

This MTA material, discovered by Mahmoud Torabinejad at Loma Linda University, was the first member of the calcium silicate family and led to a paradigm shift in the application of biomaterials in endodontics. It was commercially introduced initially in 1995 as gray ProRoot MTA, and subsequently in 1998 white MTA became available (Bansal et al., 2020; Zafar, Jamal, and Ghafoor, 2020). It is derived from refined Portland cement (75%) and contains tricalcium silicate, tricalcium aluminate, tricalcium oxides, silicate oxide, and other material oxides, forming a hydrophilic powder which sets in the presence of water (Baroudi and Samir, 2016, Kouzmanova et al., 2015; Solanki, Venkappa, and Shah, 2018). This cement has been extensively studied and has been proven to possess good physicochemical properties and to be biocompatible since the early 1990s (Parirokh and Torabinejad, 2018; Torabinejad, Parirokh, and Dummer, 2018).

MTA-Angelus

In order to minimise the long setting time and other limitations of ProRoot MTA, its composition was modified (80% PC), and MTA-Angelus was introduced in 2001. It was produced by removing dehydrated calcium sulfate (gypsum) from its composition to reduce the setting time to 10-15 minutes. Later, a fast-setting nanoparticulate white MTA was also introduced (Baroudi and Samir, 2016; Makhlouf et al., 2020).

The cements of the first two generations are already considered conventional (Baroudi and Samir, 2016; Zafar, Jamal, and Ghafoor, 2020). It was found that the white forms of both materials exhibited higher solubility than the gray ones (Borges et al., 2010; Ilić et al., 2019; Garcia et al., 2014). The gray form of ProRoot MTA is no longer available on the market.

Along with its advantages, the literature extensively discusses the drawbacks of MTA. The main disadvantages of this material are its discoloration potential, the presence of toxic elements in the material composition, difficult handling characteristics, long setting time, high material cost, and limited retreatability (Parirokh and Torabinejad, 2010b; Torabinejad, Parirokh, and Dummer, 2018).

Calcium Phosphate Based Bioceramics

In an attempt to overcome the MTA shortcomings, new improved calcium silicate-based cements with similar chemical composition and indications has been developed. Recently, a class of materials known as "Bioceramics" have been introduced and adopted. These materials represent mixtures mainly of tricalcium silicate (Ca_3SiO_5) and phosphates (Birage et al., 2019; Primus, Tay, and Niu, 2019; Zafar, Jamal, and Ghafoor, 2020). They are claimed to possess excellent biocompatibily due to their similarity to biological hydroxyapatite, intrinsic osteoinductive capacity, ability to achieve an excellent hermetic seal and to form a chemical bond with the tooth structure. (Aboud et al., 2021; Hinata et al., 2017; Raghavendra et al., 2017; Song et al., 2021). The main representatives of this group of biomaterials (third generation) are Biodentine and Bioaggregate.

Biodentine

Biodentine was developed on the basis of a unique Active Biosilicate Technology as a new class of dental material and is advertised as a "dentine substitute." It has been available on the market since 2009 (Bansal et al., 2020; Malkondu, Karapinar Kazandağ and Kazazoğlu, 2014). Unlike other bioceramics, its liquid consists of calcium chloride and a hydrosoluble polycarboxylate-based polymer that reduces the water-to-cement ratio and maximizes its strength. Its setting time is approximately 12 minutes (Makhlouf et al., 2020).

It is reported that Biodentine and MTA have shown comparable cell biocompatibility and promoted equal periradicular bone healing (Bansal et al., 2020; Tang et al., 2019). It is claimed by many authors that Biodentine is superior to other CSCs in terms of sealing ability (Bansal et al., 2020; Bolhari et al., 2015; Makhlouf et al., 2020; Shivakumar et al., 2021; Tang et al., 2019). Additionally, it poses a low risk of tooth discoloration (Tang et al., 2019) and its consistency ensures improved handling, which is better suited to clinical use than MTA (Bansal et al., 2020; Mitthra et al., 2020).

There is a plethora of reports in the literature on the successful clinical application of Biodentine as pulp capping, orthograde apical plug, perforation repair material, etc. (Allazzam et al., 2015; Anta et al., 2022; Çelik et al., 2019; Ilić et al., 2019; Jayakumar, 2018; Kouzmanova, Dimitrova, and Tsanova-Tosheva, 2021; Pradelle-Plasse et al., 2020). On the other hand, some authors reported significantly higher solubility (Singh et al., 2015; Kouzmanova and Dimitrova, 2020; Kumari et al., 2018), lower radiopacity (Camilleri et al., 2022; Ha et al., 2017; Kaur et al., 2017) and a decrease in the microhardness over time (Kouzmanova and Dimitrova, 2014) for Biodentine in comparison with MTA and other CSCs.

BioAggregate

The term "bioceramics" was originally used for BioAggregate. According to the manufacturer, this aluminum-free ceramic biomaterial was developed as a result of utilizing nano-technology (Birage et al., 2019). It is similar to MTA with respect to sealing ability and biocompatibility, but is considered to have more potential to promote mineralized tissue formation and precipitation of apatite crystals (Ha et al., 2017; Hegde, Varghese, and Malhotra, 2017; Madfa et al., 2014). Its bioactivity is comparable with MTA, but its poorer mechanical properties and long setting time limit the situations in which it could replace MTA (Mitthra et al., 2020; Zafar, Jamal, and Ghafoor, 2020). Only two retrospective clinical studies reporting the successful application of Bioaggregate as a furcal perforation repair were found in the literature (Aminov et al., 2012; Kouzmanova, Dimitrova, and Tsanova-Tosheva, 2021).

Aureoseal

Only a few studies on the physical and chemical properties of Aureoseal were found. It is a modified tetracilicate material based on Portland cement and is available only in gray form (Parirokh and Torabinejad, 2018). According to Guiliani et al., (2010) Aureoseal showed lower hardness values in an acid environment than white MTA. Its antimicrobial activity was superior to that of ProRoot MTA and calcium hydroxide (Campanella et al., 2019). Although biocompatible on bone-like cells, pulp and periodontal fibroblasts (Di Caprio et al., 2008), a moderate cytotoxic effects were reported when tested on peripheral blood monocytes (Nocca et al., 2012). Neither retrospective studies nor case reports regarding its clinical applications were found. Meanwhile, a new modification of this cement, namely Aureoseal Plus, was proposed (Cianconi et al., 2016).

In the field of endodontics, the achievement of a three-dimensional hermetic seal is one of the fundamental requirements for the endodontic materials, and especially those used for root perforation repair. This is one of the major factors associated with the healing outcome. In general, it is defined as a complex result of some qualitative properties of the applied material such as marginal adaptation, sealing ability, adhesion, solubility, volume changes, and, to some extent, microhardness (Kouzmanova, 2019; Dioguardi et al., 2021; Moustafa, Motawea, and Bayoumi, 2020).

The latter two factors have a relatively minor contribution to the effective sealing. Unfortunately, the dimensional changes of MTA have not been studied well enough yet (Ha et al., 2017), but one piece of research has shown that the volumetric changes of Biodentine were significantly lower in the acidic environment than in the saline environment, while no statistically significant difference is observed in the volume ratio of ProRoot MTA (Kwon and Seo, 2020).

Whereas the seal should be as durable and predictable as possible over time, the solubility of the material may also affect it (Moustafa, Motawea, and Bayoumi, 2020). According to the accepted standards, sealers should not be more soluble than 3% by mass (Ha et al., 2017). According to many authors, CSCs exhibit very low solubility that may be considered negligible and are practically almost insoluble (Camilleri et al., 2022; Kumari et al., 2018; Saghiri et al., 2014), but according to other investigations, some of them showed solubility higher than 3% (Camilleri et al., 2022; Kouzmanova and Dimitrova, 2020).

Therefore, the hermetic seal in the dentin-material interface must be guaranteed mainly by the good sealing capacity of the used material. The gap width between the material and dentin walls, as well as the fluid leakage depth or volume, constitutes a quantitative manifestation of its sealability (Baroudi and Samir, 2016; Kouzmanova, 2019; Dioguardi et al., 2021). The research interests are focused mainly on these two objectives.

The newer and improved calcium silicate-based materials could be seen as a promising alternative to MTA. It is believed that they overcome the drawbacks of MTA (Baroudi and Samir, 2016, Parirokh and Torabinejad, 2018; Primus, Tay, and Niu, 2019). There are numerous CSCs on the dental market and the manufacturers claim their superiority over MTA in terms of setting time, handling properties, discoloration potential, biocompatibility, and mineralization potential. Each type of cement has its own advantages and disadvantages. However, there are only a limited number of evidence-based comparative investigations regarding their efficacy as repair materials. For many of them, there is a lack of scientific information or it is too scarce.

The sealing ability of MTA and CSCs used as retrograde filling or apical plug has been investigated in detail. The hermeticity of sealing in a furcation area is less studied and comparative data are controversial. Additionally, there are some contradictions and disagreements about the dentin-cement interface characteristics and the accuracy and reproducibility of the methods used for microleakage assessment. So, on the basis of available information, it is not possible to conclude which CSCs exhibit superior sealing ability and are more suitable for perforation repair.

ProRoot MTA is now considered as a gold standard in investigations of the newly introduced hydraulic cements (Solanki, Venkappa, and Shah, 2018). Biodentine versus ProRoot MTA is the most frequently compared couple of materials in the studies (Bansal et al., 2020; Gandolfi et al., 2013; Kaur et al., 2017; Torabinejad, Parirokh, and Dummer, 2018). There is a paucity of literature regarding the sealability of other new CSCs. More studies are required to compare this property of MTA with these new materials.

Information Withholding Hypothesis

The sealing ability is a key characteristic of perforation repair materials that determines the successful clinical outcome (Baroudi and Samir, 2016). Among the different types of perforations, those located at the level of radicular furcation are considered to have a questionable and even worse prognosis (Estrela et al., 2018; Nandakumar and Nasim, 2017). Inadequate sealing of this area can result in fluid movements into the perforation defect, favouring an interradicular chronic inflammatory reaction and thus compromising the healing process (Baroudi and Samir, 2016). Microleakage is the most important risk factor, ultimately leading to a failure of the endodontic treatment (Jafari and Jafari, 2017).

None of the older materials are capable of sealing large defects and restoring normal tissue architecture at the furcal level (Baroudi and Samir, 2016; Makhlouf et al., 2020; Mitthra et al., 2020). The precise choice of a suitable repair material with the most favorable sealing properties is mandatory to ensure the longevity of the affected tooth. For this reason, the main emphasis in this study was on the evaluation of the two major aspects of the hermetic seal ability of various calcium-silicate cements from different generations used in large furcal perforation repair.

Based on the differences in the chemical formulations of the CSCs and their behavior when applied as orthograde apical plug or retrograde filling materiasl, the following *null hypotheses* were proposed: (1) The *first null hypothesis* stated that there was a significant difference in the furcation microleakage extent with the conventional MTA materials (ProRoot MTA as gold standard and MTA-Angelus) compared to the newer CSCs (Aureoseal, Bioaggregate and Biodentine) as representative materials from the Bioceramics group; (2) The *second null hypothesis* stated that there was a significant difference in the gap width values produced by the five repair materials; (3) The *third null hypothesis* stated that there was a direct correlation between these two parameters and the results obtained by both testing methodologies.

Thus, *the aim* of the present study was to evaluate and compare the hermeticity of sealing of conventional MTA products against newer calcium silicate-based cements, used as large furcal perforation repair materials, in terms of microleakage and marginal adaptation, as well as to analyze the possible correlation between these two aspects of sealability.

Method

The present study was performed in the Experimental laboratory of the Faculty of Dental Medicine,Sofia Medical University, Bulgaria and in the Institute of Mineralogy of the Bulgarian Academy of Sciences. Two conventional hydraulic MTA cements were tested as a gold standard (ProRoot MTA and MTA-Angelus) along with three newer commercially available tricalcium silicate-based materials including Aureoseal, Bioaggregate, and Biodentine.

Sample Construction

A total of 50 freshly extracted human permanent mandibular molars were used in this study, obtained from a Department of Oral and Maxillofacial Surgery. The teeth were extracted for a reason other than this study. The *inclusion criteria* were as follows: (1) molars with well-defined and sound furcation, (2) an intact crown or with minimal restorations (3) an intact pulp chamber. *Exclusion criteria:* (1) teeth with root fusion or a too narrow furcation entranc; (2) teeth with caries lesions; (3) teeth with prior root canal treatment.

The teeth were fixed in 10% buffered formalin for 2 weeks after extraction. Then they were cleaned using an ultrasonic scaler and Gracey curettes (Hu-Friedy, Leimen, Germany) to remove residual soft tissue debris and calculus, and afterwards were immersed in a mixture of 2.6% sodium hypochlorite and 3% hydrogen peroxide for 24 hours. Finally, they were rinsed under tap water and stored in physiological saline with a few thymol crystals before they were processed.

Teeth Preparation

Molars were decoronated 3 mm above the cementoenamel junction using a slow-speed diamond disc (PHM, Plovdiv, Bulgaria) under constant water spray. A standardized endodontic access cavity was made in each tooth using a high-speed Endo access diamond bur followed by an Endo-Z bur (Dentsply Sirona). The root canals were cleaned and shaped up to #40 K-files (Beutelrock, VDW GmbH, München, Germany), using a step-back technique and then filled with Cortisomol (Produits Dentaires Pierre Roland, Merignac, France) as a sealer and gutta-percha (Meta, Korea), using a single cone technique.

According to the tested material used, the molars were randomly divided into five equal experimental groups of 10 teeth each. Six additional teeth were included: three with intact pulp chamber floors and three with unrepaired perforations, which served as *negative and positive control* groups. Initially, in order to standardize the specimens in all groups, the thickness of the hard dental tissues in the furcation area of the molars was measured and the mean value for each group was calculated and compared.

Preparation of Perforations

Standardized artificial perforations were performed in the furcation area perpendicular to the center of the pulp chamber floor. Stainless steel round burs ISO 010, 012, and 014 (Komet, Gebr. Brasseler, Germany) were gradually used, in order to centralize the perforations exactly between the two roots and to mimic large defects. The burs were replaced after every five perforations. The size of the perforations was the same as the diameter of the bur, i.e., 1.4 mm, while the depth depended on the thickness of the furcation area.

In an attempt to simulate the real clinical conditions of the periodontal environment adjacent to the perforation site a clinically oriented *in vitro* approach was applied. A piece of 4 cm x 2 cm gauze was placed in the furcation area so that its ends stayed free in occlusal direction. C-silicone impression material Zetaplus Putty (Zhermack SpA, Badia Polesine, Italy) was used to provide a simulation of the tooth socket`s bony walls. The tooth roots were placed into the unset silicone up to the CEJ and then removed after the setting of the material. Thereafter, the artificial "sockets" thus formed were filled with Oranwash L Light body (Zhermack SpA, Badia Polesine, Italy) to mimic the periodontal ligament density, and the teeth were inserted again into them. The apical surfaces of the perforations were kept wet by moisturizing the free ends of the gauze with physiological saline. Furthermore, this model prevented materials from over-extrusion in the furcation area and allowed us to work without another internal matrix.

Repair of Perforations

The perforation site was cleaned with sodium hypochlorite 2.6% (Dentsply, St. Quentin en Yvelines, France), dried and sealed with different calcium-silicate cements as follows:

- Group 1 - MTA-Angelus white (Angelus, Londrina, Brazil)
- Group 2 - ProRoot MTA white (Dentsply, Tulsa, Johnson City, TN)
- Group 3 - Aureoseal (G. Ogna e Figli, Muggiò, Italia)
- Group 4 - BioAggregate (IBC Inc., Vancouver, Canada)
- Group 5 - Biodentine (Septodont, Saint-Maur-des-Fossés, France)

All materials were mixed according to the manufacturer`s instructions on a glass slab with a cement spatula to produce homogeneous paste except Biodentine, which was prepared in an amalgamator. Five drops of liquid were added to each predosed capsule of Biodentine that was placed on a capsule mixer GC Silver Mix (Dubai Medical Equipments, UAE) at a speed of 4000 rotations/min for 30 seconds. All cements were inserted into the perforation site in increments by using the Dovgan MTA carrier 0.99 mm (Quality Aspirators, Duncanwille, USA) and lightly compacted with hand pluggers (Dentsply Sirona) to the level of the pulp chamber floor. Binocular magnifying lenses Labo-clip (Eschenbach, Nürnberg, Germany) with ×2.5 magnification were used by the operator.

Moist cotton pellets were placed over the materials for 24 h. The coronal access was sealed with temporary filling material Coltosol F (Coltene,

Whaledent, Switzerland). The teeth were afterwards retrieved from the silicone model and stored in an incubator in 100% humidity at 37°C for 24 hours. The coronal access cavities of all teeth were sealed with chemically cured composite resin material Compolux (Septodont, Saint-Maur-des-Fossés, France) as a definitive restoration. All procedures were performed by one researcher.

Microleakage Assessment

The teeth were covered completely with two layers of nail varnish except for the external surface of the perforation site, so that the dye would penetrate only through the furcation area. Subsequently they were immersed for 72 hours in a 0.2% Rhodamine B dye (Alfa Aesar GmbH&Co KG, Karlsruhe, Germany). Then they were rinsed under tap water for 30 min, dried and sectioned longitudinally in a mesiodistal direction through the center of the material using a slow-speed diamond-coated disc (PHM, Plovdiv, Bulgaria) under copious amounts of water spray irrigation.

The specimens were observed, evaluated and photographed under a stereomicroscope Leica SG (Leica Microsystems, Wetzlar, Germany) at magnification x30 and the depth of linear dye penetration was measured to the nearest 0.0x mm. Four measurements were made in each sample: the length of the right and left dentinal walls of the perforation defect, and bilateral linear dye penetration in the dentin-cement interface on each wall. Afterwards, the percentage of dye penetration for each wall was calculated, and the higher value was chosen as representative.

Scanning Electron Microscopy (SEM)

Thereafter, the same specimens were gold sputtered and examined under Philips SEM 515 (Philips, Eindhoven, Netherlands) at an accelerating voltage of 25 Kv at different magnifications – from x25 to x6000. The magnification of x2000 was chosen to be the most suitable for measuring the gap size in the dentin-material interface.

The images obtained at magnification x40 were divided mentally into 2 equal parts, and four symmetrical points were chosen and magnified at x2000. The extent of the gap (distance between the dentine wall and cement) was evaluated and measured to the nearest 0.01 μm at four other equidistant points on these new micrographs by relevant software. As a result, 16 measurements for each sample were made, i.e., 160 measurements per group, and the mean gap size for all material was calculated. Finally, the possible correlation between the data obtained through the two testing methodologies was analyzed.

Statistical Analysis

All data were entered and processed with the statistical package IBM SPSS Statistics 25.0 (SPSS, Inc. Chikago, IL). Descriptive statistics (mean, standard error, standard deviation, and confidence interval) were used to evaluate the gap size and microleakage of the materials. The results were also submitted to the following tests: (1) nonparametric test of Shapiro-Wilk to check the normality of the distribution of the data; (2) test of Levene to check the homogeneity of the variance; (3) one-way ANOVA and nonparametric test of Kruskal-Wallis to evaluate differences between the groups; (4) Mann-Whithey U test for comparison between two independent samples; (5) correlation analysis (Spearman's rho) to evaluate the correlation between the gap size and dye leakage depth in all groups; (6) graphical analysis to visualize the results. α-level was set at 0.05 and P-value less than 0.05 ($p < 0.05$) was considered statistically significant.

Results

Sealing Ability (Linear Dye Penetration)

ProRoot MTA, Aureoseal, and Bioaggregate showed excellent and equivalent sealing ability without any dye penetration. Biodentine took an intermediate position with a mean value of 0.21 ± 0.63 mm, although there was no significant difference between it and the first 3 cements. A maximum dye penetration depth of 1.16 ± 1.38 mm was observed with white MTA-Angelus. On one hand, there was a statistically significant difference ($p < 0.05$) between it and materials with zero leakage, but on the other hand, concerning Biodentine such a difference did not exist in between. So, in terms of microleakage:

ProRoot MTA = Aureoseal = Bioaggregate < Biodentine < wMTA-Angelus

Descriptive statistical data are reported in Table 2 and representative stereomicroscopic pictures for all materials are shown in Figures 1-5. Additionally, when calculated as a mean percentage, the results demonstrated that the dye percolation in the wMTA-Angelus group engages more than one third of the material thickness and in the Biodentine group, approximately 1/10. A total leakage of 100% was observed in three specimens of wMTA-Angelus. The comparison of microleakage percentage relative to the length of the dentinal wall for different groups is reported in Table 2.

Table 1. Mean values of linear dye penetration in mm (Descriptive statistics)

Groups	Materials	N	Mean	SEM	SD	Min	Max
1	MTA-Angelus white	10	1.16bc	0.43	1.38	0.00	3.17
2	ProRoot MTA white	10	0.00a	0.00	0.00	0.00	0.00
3	Aureoseal	10	0.00a	0.00	0.00	0.00	0.00
4	Bioaggregate	10	0.00a	0.00	0.00	0.00	0.00
5	Biodentine	10	0.21ac	0.12	0.36	0.00	0.42

Significant variables at levels of significance ($p < 0.05$) are respectively expressed by different letters. Legend: SEM - Standard Error of the Mean, SD – Standard Deviation.

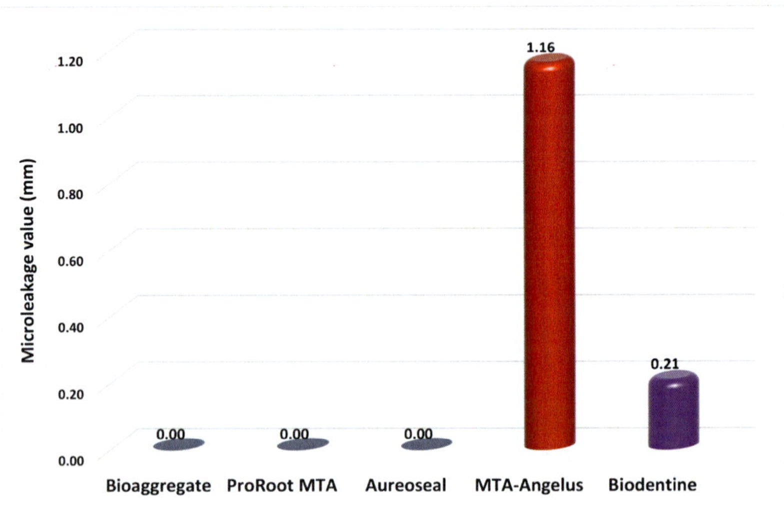

Figure 1. Comparison of the sealing ability (linear depth of dye penetration) of five calcium-silicate cements.

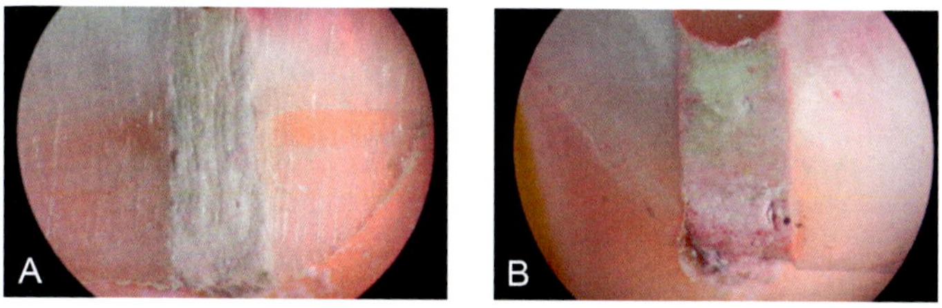

Figure 2. Microleakage in the wMTA-Angelus group seen under stereomicroscope (x30): (A) a sample with no dye penetration; (B) a sample with dye percolation.

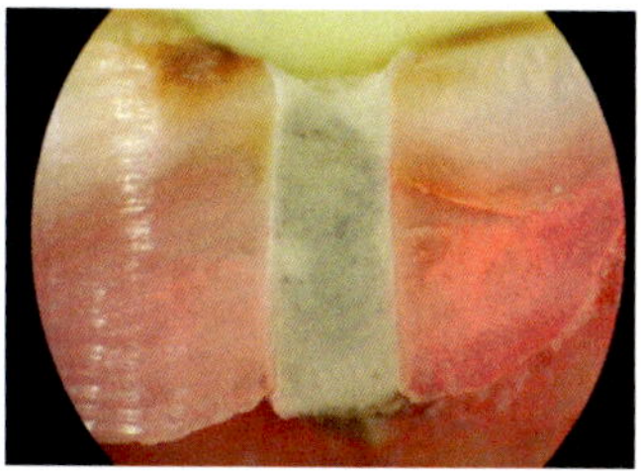

Figure 3. A representative sample for wProRoot MTA group showing a lack of dye penetration.

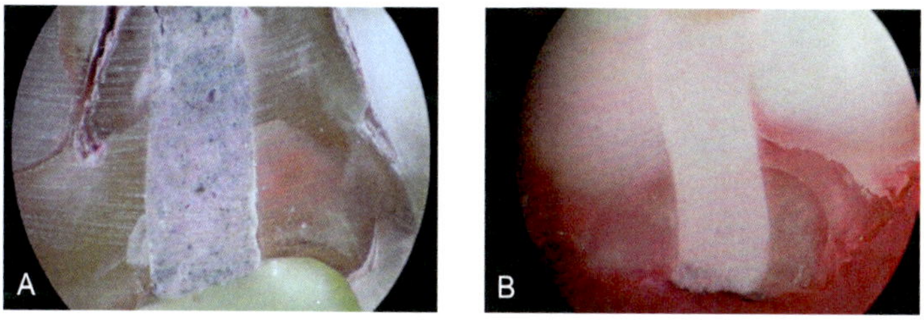

Figure 4. Lack of microleakage in representative samples for two of the bioceramics: (A) Aureoseal; (B) Bioaggregate.

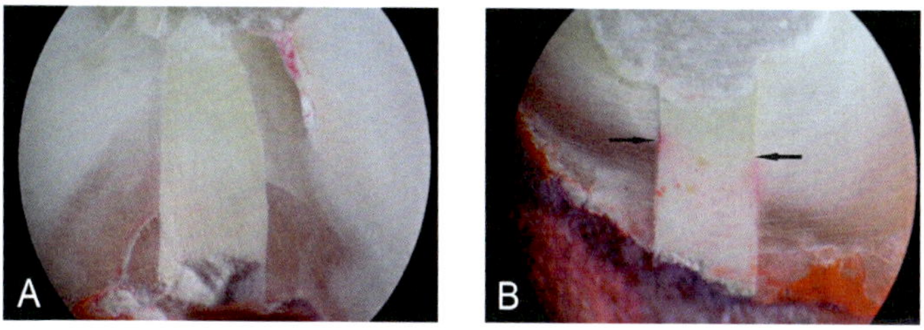

Figure 5. Microleakage in the Biodentine group: (A) a sample with no dye penetration; (B) a sample with dye leakage.

Table 2. Mean percentage of dye penetration relative to the dentinal wall length

Groups	Materials	N	Mean	SEM	SD	Min	Max
1	MTA-Angelus white	10	34.20bc	15.24	47.22	0.00	100.00
2	ProRoot MTA white	10	0.00a	0.00	0.00	0.00	0.00
3	Aureoseal	10	0.00a	0.00	0.00	0.00	0.00
4	Bioaggregate	10	0.00a	0.00	0.00	0.00	0.00
5	Biodentine	10	8.13ac	4.02	13.24	0.00	36.71

Significant variables at levels of significance ($p<0.05$) are respectively expressed by different letters. Legend: SEM - Standard Error of the Mean; SD - Standard Deviation.

Marginal Adaptation (Gap Width)

The SEM examination of the samples' longitudinal sections showed various gap size between the materials and dentinal walls across the different groups. There was no sample without a gap. The mean value and standard deviation of the marginal gap width in each group as well as the statistical differences between them are summarized in Table 3. Figure 6A is a representative picture of a perforation site filled with various cements (x40), and the 4 points are indicated on this image where new micrographs were made at magnification x2000. Figure 6B reveals one of these micrographs where the measurements are made at other four points.

Table 3. Gap size produced in each group in μm (Descriptive statistics)

No	Materials	N	Mean	SEM	95% CI	SD	Min	Max
1	MTA-Angelus white	10	6.87a	0.58	5.56-8.18	1.83	3.72	10.17
2	ProRoot MTA white	10	5.89a	0.80	4.07-7.71	2.54	2.90	10.02
3	Aureoseal	10	6.25a	0.84	4.34-8.15	2.66	2.99	11.83
4	Bioaggregate	10	5.62a	0.50	4.49-6.75	1.57	2.90	7.64
5	Biodentine	10	7.28a	0.65	5.79-8.76	2.07	3.98	9.94

Same letter reveals no significant difference at levels of significance ($p < 0.05$). Legend: SEM - Standard Error in the Mean, CI - Confident Interval, SD - Standard Deviation.

The confident intervals of the five calcium-silicate cements were very similar and these materials showed a gap size ranging from 5.62-7.28 μm as shown in Figure 7. BioAggregate produced the smallest gap size amongst them (5.62 ± 1.57 μm), whereas Biodentine had the largest gap (7.28 ± 2.07 μм). The mean gap size value increased in the following order:

BioAggregate < wProRoot < Aureoseal < wAngelus < Biodentine.

Multiple comparison tests were made to determine statistical differences between the groups but such differences were no found (p > 0.05).

Representative SEM photomicrographs for the gap width variables in different groups are shown in Figures 8-10.

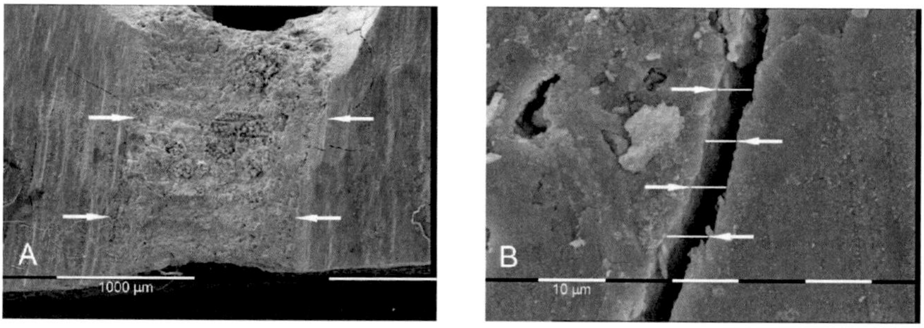

Figure 6. Representative SEM photomicrographs: (A) perforation site filled with various cements (magnification x40); (B) measurements made at four points indicated by arrows (x2000).

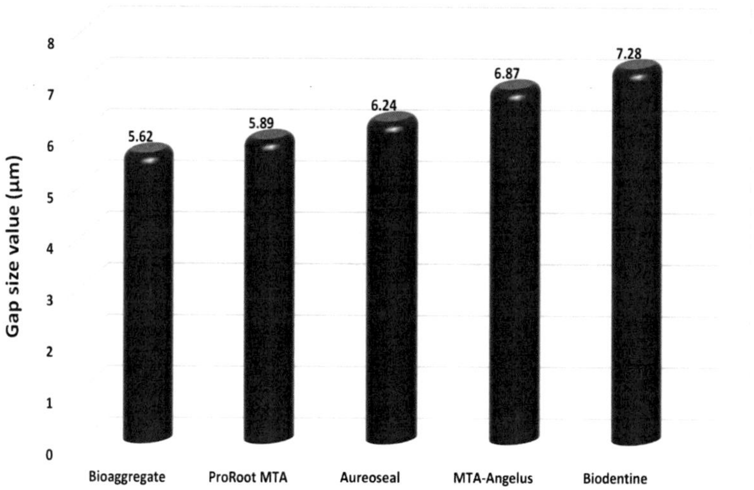

Figure 7. Comparison of the marginal adaptation (gap width) of five calcium-silicate cements.

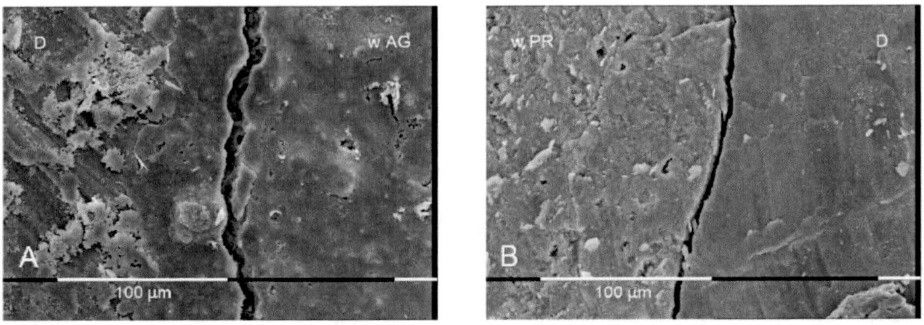

Figure 8. SEM photomicrographs (x500) of gap width produced by conventional MTA cements: (A) wMTA-Angelus; (B) wProRoot MTA.

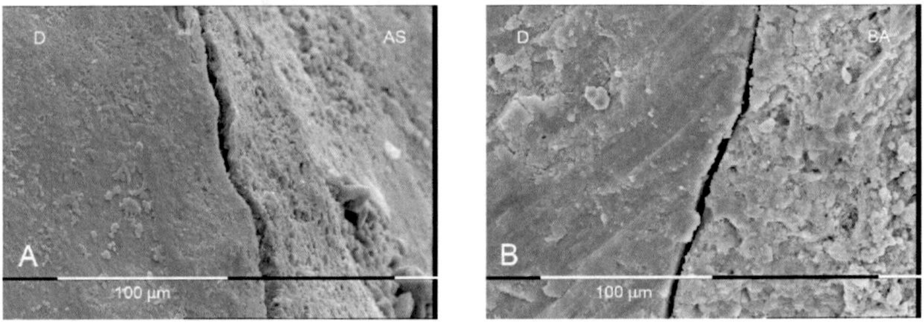

Figure 9. Photomicrographs (x500) revealing the gap size produced by two bioceramic cements: (A) Aureoseal; (B) Bioaggregate.

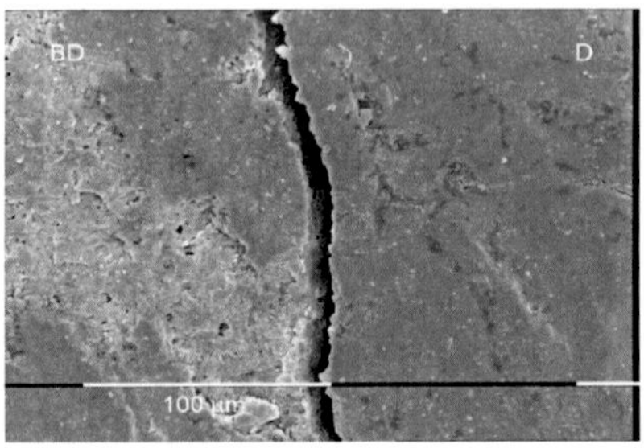

Figure 10. Representative photomicrograph (x500) of the gap size produced by Biodentine.

Correlation between the Marginal Adaptation and Sealing Ability

A correlation analysis (Spearman's rho) was used to check the relationship between two variables, namely linear dye penetration and gap width, produced by different materials. A direct correlation between the marginal adaptation and sealing ability was found, which suggests that an increase in gap width will be accompanied by an increase in microleakage. The analysis revealed a clear-cut relationship between the two aspects of the hermeticity of sealing, apparent in all groups as shown in Table 4. It was most pronounced in the wMTA-Angelus group and is graphically represented in Figure 11.

Table 4. Spearman's Linear Correlation coefficients between the gap size values and the microleakage values

No.	Materials	Mean (SD) of gap width value (μm)	Mean (SD) of leakage value (mm)	Correlation coefficients
1	Bioaggregate	5.62a (±1.57)	0.00a (±0.00)	.
2	ProRoot MTA white	5.89a (±2.54)	0.00a (±0.00)	.
3	Aureoseal	6.25a (±2.66)	0.00a (±0.00)	.
4	MTA-Angelus white	6.87a (±1.83)	1.16bc (±1.38)	0.899
5	Biodentine	7.28a (±2.07)	0.21ac (±0.63)	0.450

Significant variables at levels of significance ($p < 0.05$) are respectively expressed by different letters. Numbers in brackets are standard deviations.

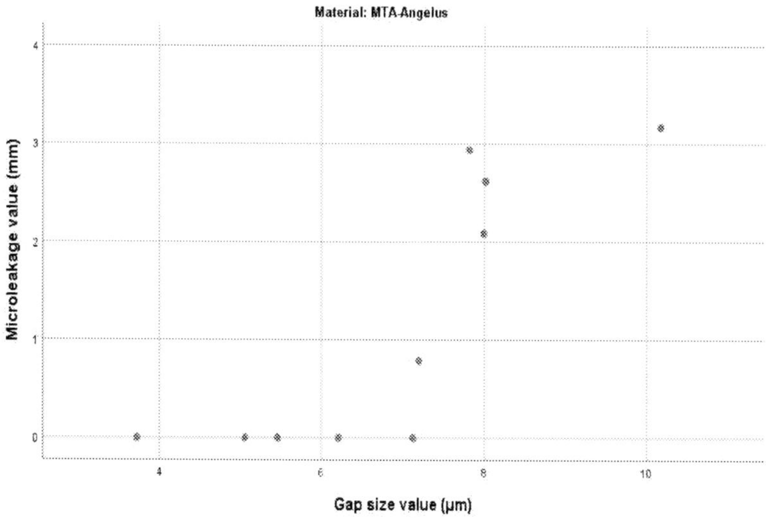

Figure 11. A diagram showing the correlation between the marginal adaptation and the sealing ability in the wMTA-Angelus group.

Table 5. Mean thickness of hard tissues in the furcation area (in mm)

Groups	Materials	N	Mean	SEM	SD	Min	Max
1	MTA-Angelus white	10	2.68a	0.16	0.51	2.35	3.55
2	ProRoot MTA white	10	2.76a	0.17	0.55	2.10	3.25
3	Aureoseal	10	3.02a	0.28	0.89	2.25	3.60
4	Bioaggregate	10	2.97a	0.27	0.87	2.05	3.65
5	Biodentine	10	2.91a	0.21	0.65	2.15	4.05

Same letter reveals no significant difference at levels of significance ($p<0.05$). Legend: SEM - Standard Error in the Mean, CI - Confident Interval, SD - Standard Deviation.

In addittion, the variations in thickness of dental hard tissues in molars' furcation area in all groups are reported in Table 5. The collected teeth had a dentin-cementum thickness of 2.05-4.05 mm, and the mean values ranged from 2.68 mm to 3.02 mm. There was no significant difference between the mean furcation thickness values in the different groups ($p < 0.05$). These measurements served as evidence that the materials were tested under equivalent conditions.

Discussion

Sealing Ability (Microleakage)

The successful outcome of the perforation recovery is highly dependent on minimizing the gap size and microleakage in the dentin wall-material interface at the perforation site (Baroudi and Samir, 2016). It should be taken into account that this is of particular importance for the perforations localized in the molar furcation area, which are more prone to failure as there is an increased chance of periodontal involvement and tissue destruction (Estrela et al., 2018; Mitthra et al., 2020).

As mentioned above, in the present study, wProRoot MTA, Bioaggregate, and Aureoseal showed superior sealing ability without any dye penetration. Biodentine showed relatively slight dye penetration (0.21 mm), but the difference was statistically insignificant ($p > 0.05$). The maximum dye penetration was seen in wMTA-Angelus (1.16 mm) ($p < 0.05$), indicating that the sealing capacity of this material is significantly worse. Therefore, the first null hypothesis tested was partially accepted and rejected with regard to ProRoot MTA.

Contrary to our findings, Nikoloudaki et al. (2014) reported that perforations restored with MTA-Angelus (7.07% ± 2.56) exhibited the least microleakage when compared to Biodentine (19.86% ± 8.67) and Aquafix Portland cement (33.43% ± 11.23) with a statistically significant difference between the groups. Petrović et al., (2015) also found a lower dye penetration value for MTA-Angelus (0.54 mm) than ours, which was significantly lower or comparable to newly synthesized nanostructured biomaterials based on active calcium silicate systems and hydroxyapatite (2.00 and 0.44 mm, respectively).

The results obtained from Makhlouf et al., (2020) suggested that Biodentine possesses a higher sealing capacity (0.11 mm) than MTA Angelus (0.46 mm). The main leakage values found by them are lower than ours. They suggested that the registered leakage values and inadequate adhesion were significantly different from the theoretical values for both materials and concluded that these materials are not ideal and still need improvement to ensure a perfect seal in case of furcal perforation.

In a recent study, Grover et al., (2020) found that Biodentine has a low mean leakage score of 0.30 against MTA Repair HP which had a score of 1.10. And as they did not measure the dye penetration in mm and score 1 was set to leakage extending to ¼ of the repair material (25%), their results indicate that the mean leakage extent of Biodentine is approximately 8%. These results are in accordance with our findings. In contrast, Bhutia et al., (2020) demonstrated that Biodentine had higher dye penetration (2.21 ± 0.412) than MTA Plus (1.77 ± 0.949), and the difference between them was significant.

According to Shivakumar et al., (2021), Biodentine showed the significantly lowest dye penetration length (467.435), followed by MTA Repair HP (758.247), and the highest penetration in the GIC group (922.021). At first glance, these findings are in agreement with some of the studies mentioned above, but on the other hand, the authors pointed out that their results were expressed in millimeters and such high values are confusing and difficult to discuss.

Haghgoo, Arfa, and Asgary (2013) evaluated the microleakage of CEM cement and tooth-colored ProRoot MTA for sealing primary molar furcal perforations and reported that both materials have a similar sealing ability. Despite the small diameter of the perforations (1 mm), in contrast to our results, they found higher depth of dye penetration with ProRoot (4.41 mm). In another study, Thota et al., (2019), using the same technique, found a lack of statistically significant difference between ProRoot MTA (1.15 ± 0.08) and Biodentine (1.05 ± 0.16), and these results are consistent with ours.

Finally, in contrast to all research cited above, Johari et al., (2020) reported full dye penetration with both Biodentine and ProRoot MTA. According to the authors, this fact might indicate partial setting of the materials when subjected to dye penetration. In order to simulate closely the clinical scenario, only 6 hours were taken for both materials to achieve complete setting prior to immersion in the dye.

Only a few studies were found in the literature using the dye penetration method to assess furcal percolation. The results obtained from other researchers who have used other techniques for microleakage measurement are not directly comparable with ours, but they give comparative information about the sealing ability of different CSCs.

Balachandran and Gurucharan (2013) assessed the sealing ability of MTA through a dye absorbance method using a ultraviolet-visible spectrophotometer. MTA showed low dye absorbance values (0.19 ± 0.07379) which were not significantly different from Bioactive bone cement (0.21 ± 0.07379). In another study, conducted by Jeevani and Jeevani (2014), using the same method, namely dye extraction with 65% nitric acid, it was observed that Endosequence showed a better sealing ability when compared with Biodentine and MICRO-MEGA MTA. Furthermore, Biodentine showed the highest dye absorbance (0.49 ± 0.07379) with a significant difference between it and other materials, thus exhibiting an insufficient sealing ability.

Katge et al., (2016) reported that the mean value of the dye absorption of Biodentine was 0.024 ± 0.031 when used in perforation repair in primary molars. The value of MTA Plus was 0.031 ± 0.026, but the difference between them was statistically insignificant. It is noteworthy that the results obtained by the last two authors with regard to Biodentine by the same technique, namely a dye extraction method, are very different. The values obtained from these authors under equivalent experimental conditions are twenty times lower than those of Jeevani et al., and this controversy is very puzzling.

Kakani and Veeramachaneni (2020) assessed the microleakage of three CSCs using a protein leakage model and the least value was observed with Biodentine (0.12015 mg/ml). They concluded that Biodentine has a significantly better sealing ability compared to MTA-Angelus, which showed the highest value (0.37384 mg/ml) and EndoSequence (0.246575 mg/ml). The authors do not specify exactly which form of MTA they have used - white or gray - but their results are in agreement with ours.

In a study by Ramazani and Sadeghi (2016), the sealing ability of ProRoot MTA and Biodentine was evaluated and compared as perforation repair materials in primary molars. A bacterial microleakage model was applied and no significant difference was found in between. In a recent study by Shah et al., (2021), conducted in 2021, similar results were found. They reported no significant difference between MTA, Biodentine, and MTA CEM when analyzed using Ultra Violet Visible Spectrophotometer (dye absorbance model). A fact of interest was the finding that a newer mixture of MTA and GIC in a 2:1 ratio showed a promising sealing ability and was comparable to other CSCs.

Övsay et al., (2018) evaluated not only the effect of the type of repair material but also the impact of perforation size and time on the amount of bacterial microleakage. The turbidity of bacteria (*E. fecalis*) was observed on the 7th, 15th, 30th, and 45th days. However, a significant difference was found only in the group of small perforations, and it was reported that ProRoot MTA is more successful in leakage prevention compared to Biodentine during the first week. These results are not fully consistent with those of Ramazani and Sadeghi and Shah et al., but they are in agreement with the findings of the current study. Contrary to that, Sinkar et al., (2015) reported the least dye absorbance with Biodentine, while RetroMTA showed the highest dye absorbance values when compared with other repair materials (dye extraction method).

Using the same assessment method, in a more recent study, Koç et al., (2021) did not register a statistically significant difference in the sealing ability among MTA-Angelus, Endocem MTA and EndoSequence BioCeramic Root Repair Material, tested at 24- and 72-hour intervals. Additionally, they reported lower leakage values, obtained at 24-hour measurements, than values measured at 72-hour intervals for all tested materials. Only MTA Angelus showed significantly better sealing ability at 72 hours compared to 24 hours. The authors concluded that the increase in the leakage values can be attributed to the chemical bonding of the hydroxyapatite layer formed on the MTA surface with the collagen fibers in the dentin over time.

Sahebi et al., (2013) compared the microleakage of ProRoot MTA and CEM cement in furcal perforation repair using a fluid filtration method. They reported that CEM cement has significantly better sealing ability compared to ProRoot MTA in all determined periods of time (24, 72, and 168 hrs.). Using a dye extraction method, Patel et al., (2014) compared the sealability of gray and white MTA under the same conditions. There was no significant difference between the two forms – 0.125% ± 0.0041 (gray MTA) and 0.127% ± 0.0035 (white MTA).

In regard to the sealing ability of Bioaggregate and Aureoseal as perforation repair materials, no studies were found using a dye leakage models or other techniques. Due to the lack of scientific information, some investigations evaluating retrograde leakage were also discussed. El Sayed and Saeed (2012) evaluated and compared the sealing ability of Diadent BioAggregate as a root-end filling material versus white ProRoot MTA. The obtained results revealed that the dye leakage was significantly less in BA (0.6 mm) when compared to wMTA (0.8 mm).

In a more recent study, Gupta and Kewalramani (2021) evaluated the microleakage of 4 bioceramics as retrograde filling materials, applying spectrophotometric measurement of extracted dye absorbance. All materials exhibited some amount of microleakage. Biodentin showed the least value, followed by Bioaggregate, but a non-significant difference was seen between them. The microleakage was highest for MTA-Angelus, and this result is in agreement with ours.

Despite a thorough literature search, only a single investigation was found with respect to the sealing ability of Aureoseal. Kasaeva and Mrikaeva (2017) reported excellent sealing of the apical part and deltoid branches of the root canal filled orthogradely with Aureoseal without any dye penetration in all samples (methylene blue). Their results coincide with ours.

Based on the results of the present study, three of the materials demonstrated zero dye penetration. In some older pieces of research, similar results are reported regarding ProRoot MTA used as root-end filling material (Andelin et al., 2002; Nicholson, 2000). Similarly to the present study, other authors also reported a lack of significant difference in microleakage between the different calcium silicate-based cements due to the similar compositions of the materials (Gupta and Kewalramani, 2021; Haghgoo, Arfa, and Asgary, 2013; Katge et al., 2016).

Although introduced in the early 1990s, ProRoot MTA exhibits an equivalent or superior sealing ability when compared to some newer or experimental CSCs in terms of microleacage (Balachandran and Gurucharan, 2013; Haghgoo, Arfa, and Asgary, 2013; Haghgoo et al., 2014; Ramazani and Sadeghi, 2016), and these findings agree with ours. Moreover, in contrast to those reporting the best sealing ability for Biodentine among other bioceramics (Grover et al., 2020; Kakani and Veeramachaneni, 2020), we reported better sealing with BioAggregate and Aureoseal.

The discrepancy between our results and other controversial results may be related to the method of sample preparation and method of evaluation. Generally, this disparity may be due to the lack of standard assessment methods. Differences in the diameter of the artificial perforations also seem to have an impact on the results obtained by researchers concerning the sealing ability of CSCs. Different researchers' artificial perforation diameters range from 0.5 mm (Samuel et al., 2016) to 4 mm (Övsay, Kaptan, and Şahin, 2018). The most commonly preferred diameter by the authors is 1 mm. It can be presumed that the smaller the perforation is the better sealing can be achieved.

Only one research focusing on the relation between microleakage and perforation size was found. Övsay, Kaptan, and Şahin (2018) investigated the impact of perforation size and root repair material type on the amount of bacterial microleakage. They found that the perforation which was 2 mm in diameter exhibited less leakage compared to a 4-mm perforation. They suggested that the damage of periodontal space would be greater as the size of the perforation increased. In such a situation, the sealing ability of the repair material would be the most significant factor determining the healing outcome.

Methods for Microleakage Assessment

The microleakage values obtained from different CSCs used as perforation repair materials varied depending on the area of application and may be attributed to the diversity in the methodology used. For this reason, it is difficult to compare the microleakage of the same materials in different studies without standardized methodology.

Several different leakage testing techniques are described in the literature to assess the ability of materials to seal furcation perforation, including dye penetration, dye extraction, fluid filtration, bacterial leakage models, an air pressure method, a radioisotope method, an electrochemical method, metal solution tracers, electrochemical tests, nanoleakage tests, and reverse diffusion (Dioguardi et al., 2021; Grover et al., 2020; Jafari and Jafari. 2017; Jeevani et al., 2014; Kakani and Veeramachaneni, 2020; Koç et al., 2021, Sahebi et al., 2013). On the other hand, these methods are very divergent in their viewpoints and there is no consensus yet regarding the best methodology to be used for the leakage assessment (Jafari and Jafari, 2017).

The microleakage dye penetration model is one of the traditional methods for microleakage evaluation. Many disadvantages of this method have been discussed in the literature, such as: (a) The molecular size of the dye particles is smaller than bacteria. (b) Rather than measuring the actual total volume of the leakage, only one plane of dye penetration can be detected and measured. (c) The obtained results show static interaction rather than dynamic interaction with periradicular tissues. (d) Destruction of the experimental specimens (Baroudi and Samir, 2016).

It is considered that the chemical characteristics, pH, and low molecular size of the dye and its dissolution by the repairing material may affect the degree of dye capillary penetration and cause leakage to be over- or under-estimated (Hanghgoo, Arfa, and Asgary, 2013). Moreover, it is claimed that the dye penetration methods do not reflect clinical conditions due to the insufficient molecular weight of the dye, which prevents it from penetrating areas where bacteria are able to access. Some authors even believe that this method is not clinically significant due to these shortcomings (Övsay, Kaptan, and Şahin, 2018; Ramazani and Sadeghi, 2016). On the other hand, it was supposed that the particle size of the dye might not be an important factor in leakage studies (Jafari and Jafari, 2017).

However, this technique is still widely used by many researchers due to its ease of performance without sophisticated devices and chemicals. The advantage of the dye penetration method compared with the other methods used for leakage measurement is its simplicity and inexpensiveness, enabling a CEM evaluation to be performed on the same set of samples. This method is easily reproducible and clinically relevant for dental perforation repair materials (Bhutia et al., 2020; Grover et al., 2020; Haghhoo, Arfa, and Asgary, 2013; Makhlouf et al., 2020). In addition, Torabinejad, Watson, and Pitt Ford (1993) stated that: "When a filling material does not allow penetration of small molecules, it has the potential to prevent leakage of larger substances such as bacteria and their by-products."

This technique was selected not only for the aforementioned advantages, but also for another important reason. Although it cannot provide volumetric evaluation of material leakage, the dye penetration method provides an idea about the thickness of the cement layer necessary for an adequate seal.

According to the different authors evaluating the apical leakage of CSCs, the minimum required thickness sufficient for an adequate retrograde seal varied from 1 to 4 mm. In an older study by Valois and Costa (2004,) a significant difference was found between 2- and 3-mm-thick MTA in preventing apical leakage. A four-mm-thick layer was significantly more effective than the other thicknesses tested. In a more recent study, Rahimi et al., (2015) evaluated the dye penetration in different thicknesses of CEM cement. Their findings demonstrated that this cement has an adequate sealing ability at a 3-mm thickness.

The depth of the perforation site is a variable factor that depends on the thickness of the hard dental tissues in the furcation area. This thickness is defined as the distance between the pulp chamber floor and the external radicular surface of the furcation (Deutsch and Musikant, 2004). According to some morphometric studies, the mean calculated value of this parameter in lower molars is 2.32-2.96 mm (Deutsch and Musikant, 2004; Lokade et al., 2011; Sterrett, Pelletier, and Russell, 1996). In the current study, this thickness in the five groups was 2.47-3.02 mm. In our previous study, it was found that the mean vertical size of the furcation is 2.78±0.65 mm in lower molars and 2.91 ± 0.63 mm in maxillary molars (Kouzmanova and Dimitrova, 2012b).

Other techniques are claimed to have many advantages and higher accuracy. In addition, they eliminate the chance of any material loss and specimen destruction (Jeevani et al., 2014; Makhlouf et al., 2020; Sahebi et al., 2013). It is considered that the dye-extraction method provides the most reliable results as it measures all of the dye taken up in the root (Baroudi and Samir, 2016). The fluid filtration technique would permit a quantitative measurement of microleakage over a longitudinal time period (Kakani and Veeramachaneni, 2020). According to Camps and Pashley (2003), the dye extraction methodology gave similar results to the fluid-filtration technique as both are based on quantitative measurements of liquid passage within interfaces.

The bacterial leakage method is considered to be an improvement over the microleakage dye penetration model because clinical bacterial contamination can be simulated (Ramazani and Sadeghi, 2016). Moradi et al., (2015) compared the accuracy of fluid filtration and bacterial leakage techniques in the assessment of the apical sealing ability of MTA and CEM. They found that both techniques showed similarity in the obtained results and there was no significant difference in between. Nikhil, Jha, and Suri (2016) evaluated and compared the sealing ability of MTA using three different methods: glucose penetration, fluid filtration, and dye penetration methods. No correlation was found among the obtained results, and the authors concluded that relying on the results of apical sealing by only one method can be misleading.

There is a lot of skepticism about the values of *in vitro* leakage studies with regard to clinical significance and limitations of the results (Gomez-Filho et al., 2012). But in a systematic review concerning the methodologies of endodontic microleakage evaluation and their importance, Jafari and Jafari (2017) stated that all methods are useful if the studies are performed strictly with large sample sizes and proper control groups, and if the technique can be standardized.

Recently, Tang et al., (2019) applied a more accurate novel test model to assess the total apical leakage of CSCs used as root-end filling materials. The

50% silver nitrate solution volume and depth that infiltrated the root canal were calculated using micro-computed tomography. Based on the quantitative analysis, a larger volume and depth of $AgNo_3$ penetration were observed in the MTA group than in the Biodentine group.

Impact of the Dye Tracer

The characteristics of the dye and its chemical interactions may affect the accuracy of the results (Hanghgoo, Arfa, and Asgary, 2013). Several substances have been used as a tracer to delineate the microleakage of endodontic materials, such as 0.5-2% methylene blue, 0.2-0.5% Rhodamine B, 50% silver nitrate solution, 0.5-2% basic fuchsin, 0.5-1% black India ink, 5% eosin, etc. (Bortoluzzi et al., 2006; Grover et al., 2020; Hanghgoo, Arfa, and Asgary, 2013; Petrović et al., 2015).

Methylene blue dye is the most commonly used dye in assessing the sealability of MTA and CSCs as furcal perforation materials. The various advantages of methylene blue compared to other dyes have influenced researchers to use it as a medium. It is cost-effective, easy to manage, and strong in colour (Grover et al., 2020, Balachandran and Gurucharan, 2013; Makhlouf et al., 2020; Sinkar et al., 2015). However, in some cases, regardless of its bright color, the maximum point of penetration is difficult to observe (Karobari et al., 2021). Other drawbacks of this dye are associated with their lower molecular weight and acidity. Its penetration ability is higher than that of the bacterial toxins (Orosco et al., 2010). Nonetheless, some researchers suggest that methylene blue exhibits leakage similar to butyric acid, a microbial metabolic product with greater penetration than Indian ink (Jafari and Jafari, 2017).

Moreover, the specific properties of MTA must also be taken into account. MTA has an alkaline pH (12.5) and contains calcium oxide, which on contact with water turns into calcium hydroxide and discolors methylene blue (Parirokh and Torabinejad, 2010a). Researchers demonstrated that its optical density value decreases on contact with MTA, thereby losing its labelling ability and causing underestimation in leakage studies (Bortoluzzi et al., 2006). Therefore, alkaline dyes are more suitable for dye leakage assessment of MTA (Hanghoo, Arfa, and Asgary, 2013).

Basic *fuchsine 0.5-1%* solution buffered to neutral pH (7.4) is a less commonly used dye with a low molecular weight (Hanghoo, Arfa, and Asgary, 2013). The molecular size of basic fuchsine is about 1-2 nm, which is smaller than bacteria (0.5–1 μm) (Johari et al., 2020). It was discovered that materials with an alkaline pH, such as MTA, cannot decolorize this dye (Stefopoulos et al., 2008). *India ink 0.5-1%* is black in colour dye with a particle diameter of ≤3 μm, smaller than the diameter of the bacteria found inside the root canal spaces. It is rarely used for sealing ability assessment of CSCs applied as perforation repair materials. In some cases, the real extent of its penetration is difficult to observe (Dewi, Asrianti, and Margono, 2017).

The use of *50% silver nitrate solution* ($AgNO_3$) is more specific. After immersion in this solution for 2 hours, the specimens have to be rinsed with photographic developer. $AgNO_3$ is not affected by the high pH of MTA. Moreover, the silver grains have a high molecular weight and similar size to that of bacteria (Costa et al., 2008; Tsatsas, Meliou, and Kerezoudis, 2005). It presents greater clearness and contrast and its use allows evaluations by SEM and micro-CT (Tang et al., 2019; Xavier et al., 2005).

Currently, *Rhodamine B* dye has been suggested as a more appropriate tracer for assessing specifically hydraulic cements. It is a fluorescent dye of organic origin, compounded by a red-violet powder and is classified as a xanthenic dye (Bortoluzzi et al., 2006; Camilleri and Pitt Ford, 2008). It was elected for the purposes of the current study for the following reasons: 1. Its penetrating capacity in gaps and dentinal tubules is higher than that of methylene blue dye (Thota et al., 2019). Rhodamine B molecules are nanometric (0.1-2 μm) and their size is comparable to the size range of a number of endodontic pathogens. Due to their low molecular weight they are ideal for simulating the penetration of bacterial enzymes and toxins in microspaces between material and dentinal walls (Jafari and Jafari, 2017). 2. It is not affected by the alkaline pH of MTA (Bortoluzzi et al., 2006). 3. Its bright color is easily detectable, even in a low concentracion (Gomes-Filho et al., 2012).

Marginal Adaptation (Gap Width)

The present study found that there was no significant difference between conventional MTA products and recently introduced bioceramic cements regarding the produced gap size. Hence, the second null hypothesis tested was totally rejected. Very few studies were found evaluating marginal adaptation of CSCs as furcal perforation repair materials under SEM. Samuel et al., (2016) conducted a study in primary molars simulating perforation with a small size (d = 0.5 mm) and found a very small gap size. According to the results obtained, the gap size produced by Biodentine (0.149 ± 0.097) was significantly lower when compared to that of MTA-Angelus (0.583 ± 0.24). Furthermore, the authors described a smaller intra-molecular space in Biodentine.

Alazrag et al. (2020) applied a more sophisticated method for sample preparation using 2 mm thick horizontal root segments placed in acrylic blocks. Perforation-like cavities were drilled parallel to the root canal. According to these authors, Biodentine showed the lowest frequency distribution of gap presence (gapped = 25% and non-gapped = 75%), followed by the MTA-Angelus (gapped = 50% and non-gapped=50%) when compared to TheraCal LC (gapped = 61% and non-gapped = 39%).

Conversely, in another study by Bhutia et al., (2020) Biodentine showed comparatively worse marginal adaptation (13.5 μm) than MTA Plus (10.4 μm) and GC Fuji VII (6.5 μm). These results are significantly higher than those found in the present study with regard to Boidentine. This could be attributed to a prolonged maturation process because of the formation of a passivating trisulfate layer over the hydrating crystals of MTA.

The literature search showed no studies assessing the marginal adaptation of Bioaggregate and Aureoseal as furcal perforation repair materials. Due to the lack of information, we also included some published data concerning retrograde application of these bioceramic materials.

Bolhari et al., (2015) evaluated the marginal adaptation of 4 CSCs in the presence of normal saline and human blood. They reported no significant difference in the measured marginal gap between the 4 filling materials: ProRoot MTA (0.81 μm), CEM cement (1.13 μm), Biodentine (0.5 μm) and BioAggregate (0.66 μm). Comparing the same materials, Bansal et al., (2019) observed similar material behavior but found a significantly larger gap width, and these results are closer to ours. They reported the minimum marginal gap score of 2.44 μm in the Biodentine group, whereas a maximum gap score of 5.63 μm was observed in the ProRoot MTA group.

In another study, Baranwal et al., (2015) compared the marginal adaptation of gray and white forms of ProRoot MTA measuring not only the width of the gap (19.54 μm vs 15.88 μm) but also the length of the gap (414.28 μm vs 645.46 μm). Finally, based on the calculated overall gap values, they concluded that gMTA exhibited a better marginal adaptation than both the white form and Portland cement (26.00 μm and 608.54 μm, respectively), but without a significant difference between them. In contrast, in our recent study, it was found that gray and white forms of Portland cement produced a superior seal (2.53 ± 1.11 μм and 4.02 ± 2.16 μм, respectively), when compared to wProRoot MTA, gMTA-Angelus, and wMTA-Angelus in terms of gap width (Dimitroba and Kouzmanova, 2012a).

Ali et al., (2020) found a gap distance of 1.52 μm produced by MTA-Angelus but did not specify which form it was – gray or white. In regard to Aureoseal, only one abstract of a poster presentation was found. D'Angelo et al., (2018) reported that the gap between the apical plug and dentin wall was higher after one week, although these differences were not statistically significant. The exact gap size was not mentioned.

Concerning the latter cited studies, it should be noted that in our opinion, the area of application may probably influence the gap size. In fact, in different cases, MTA is compacted against tissues with a different density when used as root-end filling or furcal perforation repair material.

Finally, some impressive findings concerning the dentin-to-cementum interface, associated with the specific nature and properties of MTA and CSCs, have to be mentioned and discussed. In a remarkable experimental study, Sarkar et al., (2005) characterized the interactions of MTA with a synthetic tissue fluid (STF) and root dentin. As a result, they found an adherent interfacial layer on the dentin wall that resembled hydroxyapatite in composition. The authors concluded that Ca, the dominant ion released from MTA, reacts with phosphates in STF, producing hydroxyapatite. They suggested that the sealing ability, biocompatibility, and dentinogenic activity of MTA are attributed to these physicochemical reactions. In a later study, Han and Okiji (2011) found that with Biodentine, more calcium and silicon ions were absorbed into the dentin than with MTA, resulting in the formation of tag-like structures.

Moreover, Nica et al., (2020) found good dentin penetration capacity in ProRoot MTA, Biodentine, and Endosequence RRM, used as retrograde filling materials. In all analyzed samples, concentric areas of diffuse penetration of the material into the dentinal tubules were observed, surrounding almost the entire contour of the root canal preparation. Koç et al., (2021) suggested that an intermediate hydroxyapatite layer formed on the MTA surface fills the microscopic gaps between the MTA and dentin walls, and thus positively affects the sealing property of the material due to the chemical bond between the apatite crystals and the collagen in the dentine.

Such a phenomenon was not registered in the present study. Alazrag et al., (2020) found no gap evident between the materials and dentin in 39-75% of samples in the different groups of tested materials. However, they performed their investigation in a different manner. Instead of perforations in the furcation molar area, they made perforation-like cavities in horizontal root segments. Our results showed variable marginal gaps on the dentin-material interface in all specimens. No lack of gap was found, probably because a STF was not used in this study and the observation was carried out 4-5 days after the CSCs were applied.

We believe that another important detail needs to be emphasized. The artificial perforation made with a round bur is in the geometric shape of a cylinder and the repair material follows this shape. In the present study, each specimen was sectioned longitudinally through the center of the material because if the cut does not pass exactly along this line, which corresponds to the widest diameter of the cylinder, it seems that the illusion of a lack of a gap may be created due to the overlapping of dentin by material or vice versa. Torabinejad et al. (1995) suggest that the absence of gaps in samples may be due to the presence of a thin layer of MTA at its junction with dentin. Consistent with this statement, it is worthwhile to explain that in the present study, such a phenomenon was initially observed in some samples, but after the smear layer removal, the real gap width became clearly visible.

There is a consensus regarding the accuracy of the scanning electron microscope as the most valuable tool to assess the sealing ability at the repair material-dentin interface (Moustafa, Motawea, and Bayoumi, 2020). But nowadays, new non-invasive methods for gap extent assessment with higher accuracy have been introduced and accepted. Quite recently, Toia et al., (2022) used micro-computed tomography for measuring the internal voids and gap volume, thus overcoming some shortcomings of the SEM method.

In general, Biodentine, probably the most tested cement, is considered to be superior to MTA and other new bioceramic materials (Alazrag et al., 2020; Bansal et al., 2020; Bolhary et al., 2015; Grover et al., 2020; Kakani and Veeramachaneni, 2020). According to the present study, surprisingly, Biodentine showed higher values of microleakage and gap size when compared to wProRoot MTA, Bioaggregate, and Aureoseal. The hermetic seal associated with Bioaggregate could be explained by the following: (1) It contains nano-sized particles that adhere well to the dentinal walls of the perforation defects; and (2) It is hydrophilic in nature and exhibit setting expansion (El Sayed and Saeed, 2012). Due to the lack of studies on Aureoseal, its excellent results were not discussed.

In this study, only wMTA-Angelus demonstrated significantly worse behaviour in the prevention of dye percolation, and these results agree with other reports (Alazrag et al., 2020; Makhlouf et al., 2020; Kakani and Veeramachaneni, 2020). It might be related to possible differences in setting reactions and crystallization phases of the tested materials. Moreover, Hashem and Hassanien (2008) recommended MTA-Angelus be used with an internal matrix to repair large furcation perforations.

It was very difficult to compare the results of the present study with those of other cited investigations, because they were very divergent both as absolute values and as relative comparisons of the properties exhibited by the different materials. For instance, the gap width values, reported with Biodentine varied between 0.149-13.5 μm, and those with ProRoot MTA were between 0.81-15.88 μm. Overall, among all the aforementioned results reported in the literature, our findings took an intermediate position. The inconsistency in the results obtained from different studies should be attributed to different testing conditions and methodologies applied.

Correlation Analysis

As previously mentioned, the results of this study demonstrate a direct correlation between the gap size and the depth of dye penetration. They confirmed the relationship between the two methods for assessment of the hermeticity of sealing - sealing ability using a dye penetration method and marginal adaptation through scanning electron microscopy. It seems that these results canbe attributed to the properties of the used materials. Therefore, the third null hypothesis tested was accepted.

Our study is the first investigation analysing the correlation between these two parameter values produced by different CSCs used for furcal perforation repair. To the best of the authors' knowledge, no study of this kind has been published yet with respect to furcal perforation. Only Bhutia et al., (2020) evaluated these two aspects of the sealability of MTA Plus, Biodentine, and GC Fuji VII as perforation repair materials, applying the same methods as those used in the present study. The authors found that all materials behaved in the same manner in the conditions of both testing methodologies. The results obtained through dye penetration depth and gap size measurements showed obvious similarity and consistency, but they were not subjected to statistical correlation analysis.

There are only a few reports investigating the presence or lack of such a correlation between the properties of CSCs applied at other root areas. In a recent study, Ayatollahi et al., (2017) evaluated the sealing ability of MTA and CEM cement used as apical plugs. They compared the accuracy of the fluid filtration technique and SEM and found a direct relationship between the results obtained by the two methods.

In contrast, in some older studies comparing the marginal adaptation and dye penetration of MTA and other root-end filling materials, the authors reported that an absence of correlation between the two methodologies and parameters was clearly observed (Costa et al., 2008; Xavier et al., 2005). Orosco et al., (2009) also reported no correlation between the dye leakage assay and scanning electron microscopy when studying the sealing ability and marginal adaptation of gray MTA and other materials used as apical plugs. These contradictory results may be due to the lack of standardization of assessment methods or to the use of resin replicas and indirect evaluation of the gap width (Ayatolahy et al., 2017).

Final Remarks

Other properties to consider when selecting a furcal perforation repair material, in addition to the sealing capacity, include handling characteristics and the repercussion of overfilling (Camilleri et al., 2022; Torabinejad, Parirokh and Dummer, 2018). Except Biodentine, which comes in predosed capsules, other tested CSCs are manually mixed and this may be viewed as a weak point. It has to be noted that compared to other CSCs, Biodentine showed better handling properties.

Moreover, in almost all Bioaggregate samples, the dye appeared to be absorbed throughout the material without any marginal dye percolation. This phenomenon could be attributed to the microporous structure of the material and its water soluble nature prior to achieving full setting. Additionally, the alteration in the color of ProRoot MTA (more grayish) was evident.

An inherent property of calcium silicate cements is that an unintentional over-extrusion of the material beyond the limits of the perforation and into the periodontal region would not cause any adverse effects. This is due to their bioactive nature when in direct contact with human tissues (Makhlouf et al., 2020; Torabinejad and Parirokh, 2010). In contrast, Kouzmanova (2022) reported clinical complications as a result of an over-extrusion of MTA used for an invasive cervical resorption defect repair.

Despite the widespread use of MTA and Calcium silicate-based cements in clinical dentistry, Cammilleri et al., (2022) reported no specification for standard testing of hydraulic cements and no standards explicitly for testing root-end filling materials are available. Most researchers use standards that were developed for testing other materials, such as glass ionomers, root canal sealers, resin-based materials, etc. The status quo is the same for their application as furcal perforation repair materials. Since hydraulic cements are different from other endodontic materials, the norms set for the other materials may not be appropriate for them. Hence, MTA and CSCs require specific tests or variations from the standards designed especially for them.

It should be emphasized that during their clinical application, these filling materials are in direct contact with blood or other tissue fluids. Moreover, too often the perforation repair material may be exposed to an acidic environment because of the presence of periradicular inflammation. In the early 1990s, Torabinejad et al., (1994) stated that the presence or absence of blood contamination had no significant effect on the amount of dye leakage. Another previous study showed that in the MTA-repaired furcal perforations, exposures to normal saline, blood, and lidocaine had no significant effect on the MTA properties (Hanghgoo et al., 2014).

In a more recent study, Bolhari et al., (2015) confirmed that blood contamination does not affect the marginal adaptation of ProRoot MTA, CEM cement, Biodentine or BioAggregate. Contrary to that, Bortoluzzi et al., (2006) suggests that one of the disadvantages of MTA is its tendency to wash off or disintegrate in early contact with blood and other tissue fluids.

Most importantly, in an earlier study, Lee et al., (2004) found that an acidic environment of pH 5 adversely affects both the physical properties and the hydration behavior of MTA. The pH 5 specimens' microhardness was significantly weaker compared to the samples hydrated in distilled water, normal saline and pH 7. More recently, Thomas, Chandak, and Deosarkar (2017), reported that the dislodgement resistance of Biodentine and ProRoot MTA was reduced after exposure to acidic pH. Moreover, Bosaid, Aksel, and Azim (2022) found that inflammatory pH decreased the antibacterial properties of CSCs depending on the sealer type.

Ayatollahi et al., (2016) compared the marginal adaptation of ProRoot MTA and CEM cement used as apical plugs in open apex in dry, blood- and saliva-contaminated canals. The average gap in a dry canal was significantly lower than the one in a canal contaminated with blood and saliva. Some material properties, such as solubility, surface micromorphology, and elemental composition, have been shown to be dependent on the immersion medium, including water and alternative media such as simulated body fluid, Dulbecco's modified eagle medium, HBSS (Hank's balanced salt solution), and FCS (fetal calf serum) (Camilleri et al., 2022; Benezra, Wismayer, and Camilleri, 2017; Danesh et al., 2006).

Physical property tests should replicate physiological conditions, i.e., 37°C and be submerged in physiological solution. The samples also have to mimic the clinical environment more closely (Ha et al., 2017). According to the accepted standards, when samples are tested, they should not be submerged in water but rather in a buffered saline solution in order to provide physiological concentrations of ions. Furthermore, the methodology used in standard testing should be more relevant to clinical situations (Camilleri, 2011).

Studies have shown that the hydraulic nature of calcium-silicate cements in fact requires sufficient moisture for 72 hours for ideal setting to provide low solubility and desirable sealing ability (Hahghoo et al., 2014; Makhlouf et al., 2020). In order to fulfill this requirement and simulate the real oral environment, a silicone model was used in the present study, combined with a two-layer gauze pad placed in the furcation area that was kept moisturized with physiological saline during all procedures and the setting process. It was presumed that CSCs would absorb sufficient moisture for their setting from the artificial "periodontal tissues". Additionally, a wet cotton pellet was placed in the pulp chamber on the cement surface and remained untouched for 24 hours.

Finally, it should be mentioned that coronal leakage was also observed in some specimens. Thus, it is obvious that the cause of treatment failures or successes is not solely associated with the apical leakage of the repair material. In fact, the coronal seal of the endodontic cavities may contribute greatly to treatment outcomes and should be guaranteed by high quality definitive restorations with proven good sealing ability.

All laboratory research focusing on the physico-chemical properties of the materials to be used as root perforation repair aims to provide evidence to support clinical decisions. However, the direct extrapolation of the results to the clinical situation should only be undertaken with caution, as more *in vivo* and long-term clinical investigations should confirm the laboratory findings.

Conclusion

This research examined the hermetic seal ability of five calcium silicate-based cements used in large furcal perforation repair. Under the conditions of this laboratory evaluation and within its limitations, it can be concluded that ProRoot MTA is still the material of choice for the repair of large furcation perforations among other CSCs. Except for white MTA-Angelus, all other cements demonstrated good or excellent microleakage resistance. No dye penetration was registered with ProRoot MTA, Aureoseal and Bioaggregate. A minor percolation was found in the Biodentine group.

All the tested CSCs exhibited similar behaviour in terms of marginal adaptation to the dentin walls. Therefore, Aureoseal, Bioaggregate, and Biodentine could be an efficient alternative to conventional MTA products in furcal perforation repair. Considering the obtained results, it seems that there is a direct correlation between the marginal adaptation and the sealing ability of all tested CSCs. However, in order to confirm these results, further studies using more precise new methods for total microleakage and gap volume assessment, such as micro-computed tomography, may be required.

References

Abboud, K. M., Abu-seida, A., Hassanien, E. and Tawfic, H. M. (2021). Biocompatibility of NeoMTA Plus® versus MTA Angelus as delayed furcation perforation repair materials in a dog model. *BMC Oral Health*, 192.

Akbar, I. (2015). Radiographic study of the problems and failures of endodontic treatment. *International Journal of Health Sciences, Qassim University*, 111-118.

Alamoudi, R. A., Alharbi, A. H., Farie, G. A. and Fahim, O. (2019). The value of assessing case difficulty and its effect on endodontic iatrogenic errors: a retrospective cross-sectional study. *Libyan Journal of Medicine*, 1688916.

Alazrag, M. A., Abu-Seida, A. M., El-Batouty, K. M. and El Ashry, S. H. (2020). Marginal adaptation, solubility and biocompatibility of TheraCal LC compared with MTA-angelus and biodentine as a furcation perforation repair material. *BMC Oral Health*, 298.

Aldayri, B. A., Nourallah, A. W. and Badr, F. (2019). Clinical and radiographic evaluation of the healing after MTA application on mechanical furcal perforations in primary molars – part 2. *Preprints*, 2019050322.

Allazzam, S. M., Almoudi, N. M. and E Sadek El Meligy, O. A. (2015). Clinical Applications of Biodentine in Pediatric Dentistry: A Review of Literature. *Journal of Oral Hygiene & Health*, 179.

Aminov, L., Moscalu, M., Melian, A., Salceanu, M., Hamburda, T. and Vataman, M. (2012). Clinical-radiological study on te role of biostimulating materials in iatrogenic furcation lesion. *Revista Medico-chirurgicala a Societatii de Medici si Naturalisti din Iaşi*, 907-913.

Andelin, W. E., Browning, D. F., Hsu, G. H., Roland, D. D. and Torabinejad, M. (2002). Microleakage of resected MTA. *Journal of Endodontics*, 573-574.

Anta, S., Diouma, N., Ousmane, N. S., Fatou, L. B., Florence, F. and Babacar, T. (2022). Evaluation of complete pulpotomy with Biodentine on mature permanent molars with signs and symptoms of symptomatic irreversible pulpitis: 12-months follow-up. *Journal of Endodontics*, 312-319.

Ayatollahi, F., Hazeri Baqdad Abad, M., Razavi, S. H., Tabrizizadeh, M., Ayatollahi, R. and Zarebidoki, F. (2017). Evaluating the accuracy of two microleakage assessment methods for Mineral trioxide aggregate and Calcium-Enriched Mixture cement. *Iranian Endodontic Journal*, 497-501.

Bakhtiar, H., Mirzaei, H., Bagheri, M. R., Fani, N., Mashhadiabbas, F., Baghaban Eslaminejad, M., Sharifi, D., Nekoofar, M. H. and Dummer, P. (2017). Histologic tissue response to furcation perforation repair using mineral trioxide aggregate or dental pulp stem cells loaded onto treated dentin matrix or tricalcium phosphate. *Clinical Oral Investigations*, 1579-1588.

Balachandran, J. and Gurucharan. Comparison of sealing ability of bioactive bone cement, mineral trioxide aggregate and Super EBA as furcation repair materials: A dye extraction study. (2013). *Journal of Conservative Dentistry*, 247-251.

Bansal, K., Jain, A., Aggarwal, N. and Jain, A. (2020). Biodentine VS MTA: A comparitive analysis. *International Journal of Oral Health Dentistry*, 201-208.

Bansal, R., Bansal, M., Matta, M. S., Walia, S., Kaur, B. and Sharma, N. (2019). Evaluation of marginal adaptation of MTA, Biodentine, and MTA Plus as root-end filling materials—An SEM study. *Dental Journal of Advance Studies*, 6-11.

Baranwal, A. K., Paul, M. L., Mazumdar, D., Adhikari, H. D., Vyavahare, N. K. and Jhajharia, K. (2015). An *ex-vivo* comparative study of root-end marginal adaptation using grey mineral trioxide aggregate, white mineral trioxide aggregate, and Portland cement under scanning electron microscopy. *Journal of Conservative Dentistry*, 399-404.

Baroudi, K. and Samir, S. (2016). Sealing ability of MTA used in perforation repair of permanent teeth; Literature review. *The Open Dentistry Journal*, 278-286.

Bayram, H. M., Saklar, F., Bayram, E., Orucoglu, H. and Bozkurt, A. (2015). Determination of the apical sealing abilities of Mineral trioxide aggregate, Portland cement, and Bioaggregate after irrigation with different solutions. *Journal of International Oral Health*, 13-17.

Benezra, M. K., Wismayer, P. S. and Camilleri J. (2017). Influence of environment on testing of hydraulic sealers. *Scienticic Reports*, 17927.

Bhutia, S. Y., Goel, M., Kumar, V. and Verma, S. (2020). Comparative evaluation of sealing ability and marginal adaptation of furcation repair materials: An *in vitro* study. *Journal of Advances in Medicine and Medical Research*, 35-43.

Birage, P. P., Bansode, P. V., Pathak, S. D. and Wavdhane, M. B. (2019). Root perforations & its management: A review article. *International Journal of Medical Science and Advanced Clinical Research*, 69-77.

Bolhari, B., Ashofteh Yazdi, K., Sharifi, F. and Pirmoazen, S. (2015). Comparative scanning electron microscopic study of the marginal adaptation of four root-end filling materials in presence and absence of blood. *Journal of Dentistry of Tehran University of Medical Sciences*, 226-234.

Borges, A. H., Pedro, F. L., Miranda, C. E., Semenoff-Segundo, A., Pécora, J. D. and Filho, A. M. (2010). Comparative study of physico-chemical properties of MTA-based and Portland cements. *Acta Odontologica Latinoamericana*, 175-181.

Bortoluzzi, E. A., Broon, N. J., Bramante, C. M., Garcia, R. B., de Moraes, I. G. and Bernardineli, N. (2006). Sealing ability of MTA and radiopaque Portland cement with or without calcium chloride for root-end filling. *Journal of Endodontics*, 897-900.

Bosaid, F., Aksel, H. and Azim, A. A. (2022). Influence of acidic pH on antimicrobial activity of different calcium silicate based-endodontic sealers. *Clinical Oral Investigations*, Pub Date: 2022 May 5. Online ahead of print.

Camilleri, J. (2011). Evaluation of the effect of intrinsic material properties and ambient conditions on the dimensional stability of white mineral trioxide aggregate and Portland cement. *Journal of Endodontics*, 239-245.

Camilleri, J. (2020). Endodontics in the era of hydraulic materials. *ENDO Endodontic Practice Today*, 171-172.

Camilleri, J., Atmeh, A., Li, X. and Meschi, N. (2022). Present status and future directions: Hydraulic materials for endodontic use. *International Endodontic Journal*, 710-777.

Camilleri, J. and Pitt Ford, T. R. (2008). Evaluation of the effect of tracer pH on the sealing ability of glass ionomer cement and mineral trioxide aggregate. *Journal of Materials Science. Materials in Medicine*, 2941-2948.

Camilleri, J., Wang, C., Kandhari, S., Heran, J. and Shelton, R. M. (2022). Methods for testing solubility of hydraulic calcium silicate cements for root-end filling. *Scientific Reports*, 7100.

Campanella, V., Mummolo, S., Grazzini, F., Barlattani, A. and Di Girolamo, M. (2019). The effectiveness of endodontic sealers and endodontic medicaments on the elimination of *Enterococcus faecalis*: An *in vitro* study. *Journal of Biological Regulators & Homeostatic Agents*, 97-102.

Camps, J. and Pashley, D. (2003). Reliability of the dye penetration studies. *Journal of Endodontics*, 592-594.

Çelik, B. N., Mutluay, M. S., Arıkan, V. and Sarı, Ş. (2019). The evaluation of MTA and Biodentine as a pulpotomy materials for carious exposures in primary teeth. *Clinical Oral Investigations*, 661-666.

Cervino, G., Laino, L., D'Amico, C., Russo, D., Nucci, L., Amoroso, G., Gorassini, F., Tepedino, M., Terranova, A., Gambino, D., Mastroieni, R., Tözüm, M. D. and Fiorillo, L. (2020). Mineral trioxide aggregate applications in endodontics: A review. *European Journal of Dentistry*, 683-691.

Cianconi, L., Palopoli, P., Campanella, V. and Mancini, M. (2016). Composition and microstructure of MTA and Aureoseal Plus: XRF, EDS, XRD and FESEM evaluation. *European Journal of Paediatric Dentistry*, 281-285.

Costa, A. T., Post, L. K., Xavier, C. B., Weber, J. B. and Gerhardt-Oliveira, M. (2008). Marginal adaptation and microleakage of five root-end filling materials: an *in vitro* study. *Minerva Stomatologica*, 295-300.

Danesh, G., Dammaschke, T., Gerth, H. U., Zandbiglari, T. and Schäfer, E. (2006). A comparative study of selected properties of ProRoot mineral trioxide aggregate and two Portland cements. *International Endodontic Journal*, 213-219.

D'Angelo, M., Savo, F., Donfrancesco, O., De Lucenti, M., Zanza, A., Gambarini, G. and Testarelli, L. (2018). Evaluation of the accuracy of apical gap for MTA: An *in vitro* study. *III SIE International Congress. Endodontics: Clinical Solutions*. Milan, 8-10 Nov. Poster presentation No 37, Abstract book, pp. 66-67.

de Melo, P. H. A., Machado, A. G., Machado, A. L. B., Carvalho, E. N., de Melo, J. B. and Schneider, L. F. J. (2019). Evaluation of root perforation treatments with mineral trioxide aggregate: A retrospective case series study. *Iranian Endodontic Journal*, 144-151.

Deutsch, A. S. and Musikant, B. L. (2004). Morphological measurements of anatomic landmarks in human maxillary and mandibular molar pulp chambers. *Journal of Endodontics*, 388-390.

Dewi, F., Asrianti, D. and Margono, A. (2017). Microleakage evaluation of modified mineral trioxide aggregate effect toward marginal adaptation on cervical dentin perforation. *International Journal of Applied Pafmaceutics*, 10-13.

Di Caprio, M. P., Spagnuolo, G., D`Anto, V., Simeone, M., Ambrosio, L. and Rengo, S. (2008). Biocompatibility of a new Portland cement-based materials for endodontics. *PER of the IADR*, London, UK, 10-12 Sept, Poster Presentation ID: 439.

Dioguardi, M., Alovisi, M., Sovereto, D., Troiano, G., Malagnino, G., Di Cosola, M., Cazzolla, A. P., Laino, L. and Lo Muzio, L. (2021). Sealing ability and microbial leakage of root-end filling materials: MTA versus epoxy resin: A systematic review and meta-analysis. *Heliyon*, e07494.

El Sayed, M. and Saeed, M. (2012). *In vitro* comparative study of sealing ability of Diadent BioAggregate and other root-end filling materials. *Journal of Conservative Dentistry*, 249-252.

Estrela, C., de Almeida Decurcio, D., Rossi-Fedele, G., Silva, J. A., Guedes, O. A. and Borges, A. H. Root perforations: a review of diagnosis, prognosis and materials. *Brazilian Oral Research*, e73.

Gandolfi, M. G., Siboni, F., Polimeni, A., Bossù, M., Riccitiello, F., Rengo, S. and Prati, C. (2013). *In vitro* screening of the apatite-forming ability, biointeractivity and physical properties of a tricalcium silicate material for endodontics and restorative dentistry. *Dentistry Journal*, 41-60.

Garcia, L. da F. R., Chinelatti, M. A., Rossetto, H. L. and Pires-de-Souza, F. de C. (2014). Solubility and disintegration of new calcium aluminate cement (EndoBinder) containing different radiopacifying agents. *Journal of Endodontics*, 261-265.

Goldberg, M. (2020). Failures of endodontic treatment: Pulp floor and root perforations. *Acta Scientific Microbiology*, 58-64.

Giuliani, V., Nieri, M., Pace, R. and Pagavino, G. (2010). Effects of pH on surface hardness and microstructure of mineral trioxide aggregate and Aureoseal: an *in vitro* study. *Journal of Endodontics*, 1883-1886.

Gomes-Filho, J. E., Moreira, J. V., Watanabe, S., Lodi, C. S., Cintra, L. T., Dezan Junior, E., Bernabé, P. F., Nery, M. J. and Otoboni Filho, J. A. (2012). Sealability of MTA and calcium hydroxide-containing sealers. *Journal of Applied Oral Science*, 347-351.

Grover, R., Sadana, G., Gupta, S., Gupta, T., Mehra, M. and Kaur, B. (2020). Comparative evaluation of sealing ability of two different biocompatible materials in repair of furcal perforation: An *in vitro* study. *Dentistry & Medical Research*, 70-74.

Gupta, R. and Kewalramani, R. (2021). In-vitro evaluation of microleakage of bioceramic root-end filling materials: A spectrophotometric study. *Journal of Oral Biology and Craniofacial Research*, 330-333.

Ha, W. N., Nicholson, T., Kahler, B. and Walsh, L. J. (2017). Mineral trioxide aggregate-A review of properties and testing methodologies. *Materials (Basel)*, 1261.

Haghgoo, R., Arfa, S. and Asgary, S. (2013). Microleakage of Calcium Enriched Mixture and ProRoot Mineral trioxide aggregate as furcal perforation repair materials in primary teeth. *Iranian Endodontic Journal*, 187-190.

Haghgoo, R., Niyakan, M., Nazari Moghaddam, K., Asgary, S. and Mostafaloo, N. (2014). An *in vitro* comparison of furcal perforation repaired with Pro-root MTA and New Endodontic Cement in primary molar teeth- A microleakage study. *Journal of Dentistry Shiraz University of Medical Sciences*, 28-32.

Haji-Hassani, N., Bakhshi, M. and Shahabi, S. (2015). Frequency of iatrogenic errors through root canal treatment procedure in 1335 chart of dental patients. *Journal of International Oral Health*, 14-17.

Han, L. and Okiji, T. (2011). Uptake of calcium and silicon released from calcium silicate-based endodontic materials into root canal dentine. *International Endodontic Journal*, 1081-1087.

Hashem, A. A. and Hassanien, E. E. (2008). ProRoot MTA, MTA-Angelus and IRM used to repair large furcation perforations: sealability study. *Journal of Endodontics*, 59-61.

Hegde, M., Varghese, L. and Malhotra, S. (2017). Tooth perforation repair – A review. *Oral Health and Dental Management*, 1-4.

Hinata, G., Yoshiba, K., Han, L., Edanami, N., Yoshiba, N. and Okiji, T. (2017). Bioactivity and biomineralization ability of calcium silicate-based pulp-capping materials after subcutaneous implantation. *International Endodontic Journal*, e40-e51.

Ilić, D. V., Antonijević, Đ. M., Biočanin, V. M., Čolović, B., Danilović, W., Komlev, V. S., Teterina, A. Y. and Jokanović, V. R. (2019). Physico-chemical and biological properties of dental calcium silicate cements – literature review. *Hemijska Industrija*, 281-294.

Jafari, F. and Jafari, S. (2017). Importance and methodologies of endodontic microleakage studies: A systematic review. *Journal of Clinical and Experimental Dentistry*, e812-e819.

Jayakumar, M. MTA: A review of literature with clinical application. (2018). *IOSR Journal of Dental and Medical Sciences*, 4-11.

Jeevani, E., Jayaprakash, T., Bolla, N., Vemuri, S., Sunil, C. R. and Kalluru, R. S. (2014). Evaluation of sealing ability of MM-MTA, Endosequence, and biodentine as furcation repair materials: UV spectrophotometric analysis. *Journal of Conservative Dentistry*, 340-343.

Johari, Y., Pungut, N. A., Yin, V. W. X., Khamis, M. F. and Md Yusoff, M. N. (2020). Sealing ability of injectable dental composites, Biodentine and MTA in repairing furcal perforation of permanent molar teeth. *Journal of International Dental and Medical Research*, 1428-1434.

Kahler, B., Chugal, N. and Lin, L. M. (2017). Alkaline materials and regenerative endodontics: A review. *Materials (Basel)*, 1389.

Kakani, A. K. and Veeramachaneni, C. (2020). Sealing ability of three different root repair materials for furcation perforation repair: An *in vitro* study. *Journal of Conservative Dentistry*, 62-65.

Kakani, A. K., Veeramachaneni, C., Majeti, C., Tummala, M. and Khiyani, L. (2015). A review on perforation repair materials. *Journal of Clinical & Diagnostic Research*, ZE09-13.

Karobari, M. I., Basheer, S. N., Sayed, F. R., Shaikh, S., Agwan, M. A. S., Marya, A., Messina, P. and Scardina, G. A. (2021). An *in vitro* stereomicroscopic evaluation of bioactivity between Neo MTA Plus, Pro Root MTA, BIODENTINE & glass ionomer cement using dye penetration method. *Materials (Basel)*, 3159.

Kasaeva, A. I. and Mrikaeva O. M. (2017). Experimental assessment of the degree of the sealing of the dentinal tubules and the apical orifice of the root canal material Aureoseal (Ogna). *The Journal of Scientific Articles: "Health and Education Millennium"*, 29-30.

Kaur, M., Singh, H., Dhilon, J. S., Batra, M. and Saini, M. (2017). MTA versus Biodentine: Review of literature with a comparative analysis. *Journal of Clinical and Diagnostic Research*, ZG01-ZG05.

Koç, C., Aslan, B., Ulusoy, Z. and Oruçoğlu, H. (2021). Sealing ability of three different materials to repair furcation perforations using computerized fluid filtration method. *Journal of Dental Research, Dental Clinics, Dental Prospects*, 183-187.

Kouzmanova, Y. (2019). Endodontic perforations – hermeticity of sealing with calcium silicate cements, and prevention. *Thesis*, Sofia Medical University.

Kouzmanova, Y. (2022). Invasive cervical resorption due to iontophoretic internal bleaching with hydrogen peroxide: two case reports. *Journal of IMAB - Annual Proceeding (Scientific Papers)*, 4361-4365.

Kouzmanova, Y. and Dimitrova, I. (2012a). Sealing ability of three mineral trioxide aggregate products, two new calcium-cilicate cements and one resin-modified glass-ionomer cement: Comparative laboratory investigation. *Dental Medicine*, 5-13.

Kouzmanova, Y. and Dimitrova, I. (2012b). Morphometric stugy of the furcation of molars. *Dental Medicine*, 169-174.

Kouzmanova, Y. and Dimitrova, I. (2014). Microhardness of calcium silicate-based cements. Comparative investigation. *Dental Medicine*, 108-115.

Kouzmanova, Y. and Dimitrova, I. (2020). Solubility of calcium silicate-based cements. *Acta Medica Bulgarica*, 27-29.

Kouzmanova, Y. and Nikiforova, H. (2013). Microleakage of four root-end filling materials. *Dental Medicine*, 2013, 138-144.

Kouzmanova, Y. I., Dimitrova, I. V., Dentscheva, G. D., Aleksandrov, L. I., Markova-Velichkova, M. G. and D. G. Kovacheva, D. G. (2015). Comparative study of the phase formation and interaction with water of calcium-silicate cements with dental applications. *Bulgarian Chemical Communications*, 239-244.

Kouzmanova, Y. I., Dimitrova, I. V. and Tsanova-Tosheva, D. M. (2021). Long-term clinical repair of endodontic perforations treated with calcium silicate-based cements. (2021). In BECB 2021. *ACM'S International Conference Proceeding Series*, New York, USA, Article 13, 1-3.

Krupp, C., Bargholz, C., Brüsehaber, M. and Hülsmann, M. (2013). Treatment outcome after repair of root perforations with mineral trioxide aggregate: a retrospective evaluation of 90 teeth. *Journal of Endodontics*, 1364-1368.

Kumari, S., Mittal, A., Dadu, S., Dhaundival, A., Abraham, A. and Yendrembam, B. (2018). Comparative evaluation of physical and chemical properties of calcium silicate-based root-end filling materials (Mineral trioxide aggregate and biodentine): An *in vitro* study. *Indian Journal of Dental Sciences*, 197-202.

Kunjan, A. P. and Ballal, N. V. (2020). Calcium silicate based cements in endodontics. *Journal of International Dental and Medical Research*, 1183-1190.

Kwon, S. Y. and Seo, M. S. (2020). Comparative evaluation of volumetric changes of three different retrograde calcium silicate materials placed under different pH condititions. *BMC Oral Health*, 330.

Lee, S. J., Monsef, M. and Torabinejad, M. (1993). Sealing ability of a mineral trioxide aggregate for repair of lateral root perforations. *Journal of Endodontics*, 541-544.

Lee, Y. L., Lee, B. S., Lin, F. H., Lin A. Y., Lan W. H. and Lin, C. P. (2004). Effects of physiological environments on the hydration behavior of mineral trioxide aggregate. *Biomaterials*, 787-793.

Lokade, J., Rawlani, S., Baheti (Chandak), R., Roy, S., Chandak, M. and Lohe, V. (2011). Morphological measurement of anatomic landmarks in human mandibular molar pulp chamber – an *in vivo* study. *Journal of Korean Dental Science*, 1-5.

Madfa, A. A., Al-Sanabani, F. A. and Al-Qudami Al-Kudami, N. H. (2014). Endodontic repair filling materials: A review article. *Journal of Advances in Medicine and Medical Research*, 3059-3079.

Mahmood Talabani, R., Taha Garib, B. and Masaeli, R. (2020). The response of the pulp-dentine complex, PDL, and bone to three calcium silicate-based cements: A histological study in an animal rat model. *Bioinorganic Chemical Applications*, 9582165.

Makhlouf, M., Zogheib, C., Makhlouf, A.-C, Kaloustian, M. K., El Hachem, C. and Habib, M. (2020). Sealing ability of calcium silicate-based materials in the repair of furcal perforations: a laboratory comparative study. *The Journal of Contemporary Dental Practice*, 1091-1097.

Malkondu, Ö., Karapinar Kazandağ, M. and Kazazoğlu, E. (2014). A review on biodentine, a contemporary dentine replacement and repair material. *BioMed Research International*, 2014: 160951.

Mente, J., Leo, M., Panagidis, D., Saure, D. and Pfefferle, T. (2014). Treatment outcome of mineral trioxide aggregate: repair of root perforations-long-term results. *Journal of Endodontics*, 790-796.

Mitthra, S., Shobhana, R., Venkatachalam, P. and Vivekanandhan, P. (2020). An overview on root perforations: Diagnosis, prognosis and management. *European Journal of Molecular & Clinical Medicine*, 1240-1244.

Moradi, S., Lomee, M. and Gharechahi, M. (2015). Comparison of fluid filtration and bacterial leakage techniques for evaluation of microleakage in endodontics. *Dental Research Journal (Isfahan)*, 109-114.

Moustafa, A. A., Motawea, I. T. and Bayoumi, R. E. (2020). Solubility and sealing ability of recently introduced biocearmic root-end filling materials. *Al Azhar Dental Journal for Girls*, 547-557.

Nabavizadeh, M. R., Moazzami, F., Gholami, A., Mehrabi, V. and Ghahramani, Y. (2022). Cytotoxic effect of Nano Fast Cement and ProRoot Mineral trioxide aggregate on L-929 fibroblast cells: an *in vitro* study. *Journal of Dentistry (Shiraz)*, 13-19.

Nandakumar, M. and Nasim, I. (2017). Management of perforation – A review. *Journal of Advanced Pharmacy Education & Research*, 208-211.

Nica, L. M., Boscornea-Pușcu, A. S., Horhat, R. M., Karsoua, M. and Cîrligeriu, L. E. (2020). Comparison of three calcium silicate cements used as retrograge filling materials. An *in vitro* study. *Research and Clinical Medicine*, 32-36.

Nicholson, J. (2000). Mineral trioxide aggregate — a new experimental material for retrograde filling. *British Dental Journal,* 259.

Nikhil, V., Jha, P. and Suri, N. K. (2016). Effect of methods of evaluation on sealing ability of mineral trioxide aggregate apical plug. *Journal of Conservative Dentistry*, 231-234.

Nikoloudaki, G. E., Kontogiannis, T., Meliou, H. A. and Kerezoudis, N. P. (2014). A comparative *in-vitro* study of sealing ability of four different materials used in furcation perforation. *Open Journal of Stomatology,* 402-11.

Nocca, G., Lupi, A., Testarelli, L., Lupp, A., Pacifici, L., Pompa, G., Vozza, I. and Gambarini, G. (2012). Biocompatibility of root canal filling materials: differences between vitality and functionality tests. *European Journal of Inflammation*, 105-110.

Orosco, F. A., Bramante, C. M., Garcia, R. B., Bernardineli, N. and de Moraes, I. G. (2010). Sealing ability, marginal adaptation and their correlation using three root-end filling materials as apical plugs. *Journal of Applied Oral Science*, 127-34.

Övsay, E., Kaptan, R. F. and Şahin, F. (2018). The repair of furcal perforations in different diameters with Biodentine, MTA, and IRM repair materials: A laboratory study using an *E. Faecalis* leakage model. *BioMed Research International*, 5478796.

Parirokh, M. and Torabinejad, M. (2010a). Mineral trioxide aggregate: A comprehensive literature review – part I: chemical, physical, and antibacterial properties. *Journal of Endodontics*, 16-27.

Parirokh, M. and Torabinejad, M. (2010b). Mineral trioxide aggregate: A comprehensive literature review – part III: clinical application, drawbacks, and mechanism of action. *Journal of Endodontics*, 400-413.

Parirokh, M. and Torabinejad, M. (2018). Mineral trioxide aggregate and other bioactive endodontic cements: an updated overview – part I: vital pulp therapy. *International Endodontic Journal*, 177-205.

Patel, N., Patel, K., Baba, S. M., Jaiswal, S., Venkataraghavan, K. and Jani, M. (2014). Comparing gray and white mineral trioxide aggregate as a repair material for furcation perforation: an *in vitro* dye extraction study. *Journal of Clinical and Diagnostic Research*, ZC70-73.

Petrović, V., Galić, V. O., Dželetović, B., Jokanović, V. and Živković, S. (2015). Marginal microleakage of newly synthesized nanostructured biomaterials based on active calcium silicate systems and hydroxyapatite. *Serbian Dental Journal*, 109-116.

Pontius, V., Pontius, O., Braun, A., Frankenberger, R. and Roggendorf, M. J. (2013). Retrospective evaluation of perforation repair in 6 private practices. *Journal of Endodontics*, 1346-1358.

Pradelle-Plasse, N., Mocquot, C., Semennikova, K., Colon, P. and Grosgogeat, B. (2020). Interface between calcium silicate cement and adhesive systems according to adhesive families and cement maturation. *Restorative Dentistry & Endodontics*, e3.

Primus, C. M., Tay, F. R. and Niu, L. N. (2019). Bioactive tri/dicalcium silicate cements for treatment of pulpal and periapical tissues. *Acta Biomaterialia*, 35-54.

Raghavendra, S. S., Jadhav, G. R., Gathani, K. M. and Kotadia, P. (2017). Bioceramics in endodontics - A review. *Journal of Istanbul University Faculty of Dentistry*, S128-S137.

Rahimi, S., Asgary, S., Samiei, M., Bahari, M., Vahid Pakdel, S. M. and Mahmoudi, R. (2015). The effect of thickness on the sealing ability of CEM cement as a root-end filling material. *Journal of Dental Research, Dental Clinics, Dental Prospects*, 6-10.

Ramazani, N. and Sadeghi, P. (2016). Bacterial leakage of Mineral trioxide aggregate, Calcium-Enriched Mixture and Biodentine as furcation perforation repair materials in primary molars. *Iranian Endodontic Journal*, 214-218.

Saed, S. M., Ashley, M. P. and Darsey, J. (2016). Root perforations: aetiology, management strategies and outcomes. The hole truth. *British Dental Journal*, 171-180.

Saghiri, M. A., Godoy, F. G., Gutmann, J. L., Lotfi, M., Asatourian, A., Sheibani, N. and Elyasi, M. (2014). The effect of pH on solubility of nano-modified endodontic cements. *Journal of Conservative Dentistry*, 13-77.

Sahebi, S., Moazami, F., Sadat Shojaee, N. and Layeghneghad, M. K. (2013). Comparison of MTA and CEM cement microleakage in repairing furcal perforation, an *in vitro* study. *Journal of Dentistry Shiraz University of Medical Sciences*, 31-36.

Sălceanu, M., Topoliceanu, C., Ghiorghe, C. A., Aminov, L., Hamburda, T., Giuroiu, C., Dekoli, Y., Antohi, C. and Melian, A. (2017). Clinical and radiographic study regarding the success therapy of the root furcation lesions: A 11 months follow up. *Romanian Journal of Oral Rehabilitation*, 11-17.

Samuel, A., Asokan, S., Geetha Priya, P. R. and Thomas, S. (2016). Evaluation of sealing ability of Biodentine™ and mineral trioxide aggregate in primary molars using scanning electron microscope: A randomized controlled *in vitro* trial. *Contemporary Clinical Dentistry*, 322-325.

Sarao, S. K., Berlin-Broner, Y. and Levin, L. (2021). Occurrence and risk factors of dental root perforations: a systematic review. *International Dental Journal*, 96-105.

Sarkar, N. K., Caicedo, R., Ritwik, P., Moiseyeva, R. and Kawashima, I. (2005). Physicochemical basis of the biologic properties of mineral trioxide aggregate. *Journal of Endodontics*, 97-100.

Shah, R., Daudi, A. A., Nahar, P., Misar, P., Patidar, S. and Patidar, S. (2021). A Comparative evaluation of sealing ability of five different materials for furcation perforation repair using UV-Spectrophotometric analysis: An *in-vitro* study. *Mymensingh Medical Journal*, 840-845.

Shivakumar, V. H., Tegginamani, A. S., Dicksit, D. D. and Zamzuri, A. T. B. (2021). An evaluation on sealing ability of calcium silicate-based cements and glass ionomer cement as perforation repair materials, *Journal of Pharmaceutical Research International*, 137-145.

Singh, S., Podar, R., Dadu, S., Kulkarni, G. and Purba, R. Solubility of a new calcium silicate-based root-end filling material. *Journal of Conservative Dentistry*, 149-153.

Sinkar, R. C., Patil, S. S., Jogad, N. P. and Gade, V. J. (2015). Comparison of sealing ability of ProRoot MTA, RetroMTA, and Biodentine as furcation repair materials: An ultraviolet spectrophotometric analysis. *Journal of Conservative Dentistry*, 445-448.

Solanki, N. P., Venkappa, K. K. and Shah, N. C. (2018). Biocompatibility and sealing ability of mineral trioxide aggregate and biodentine as root-end filling material: A systematic review. *Journal of Conservative Dentistry*, 10-15.

Song, W., Li, S., Tang, Q., Chen, L. and Yuan, Z. (2021). *In vitro* biocompatibility and bioactivity of calcium silicate-based bioceramics in endodontics (Review). *International Journal of Molecular Medicine*, 128.

Stefopoulos, S., Tsatsas, D. V., Kerezoudis, N. P. and Eliades, G. (2008). Comparative *in vitro* study of the sealing efficiency of white vs grey ProRoot mineral trioxide aggregate formulas as apical barriers. *Dental Traumatology*, 207-213.

Sterrett, J. D., Pelletier, H. and Russell, C. M. (1996). Tooth thickness at the furcation entrance of lower molars. *Journal of Clinical Periodontology*, 621-627.

Tang, J. J., Shen, Z. S., Qin, W. and Lin, Z. (2019). A comparison of the sealing abilities between Biodentine and MTA as root-end filling materials and their effects on bone healing in dogs after periradicular surgery. *Journal of Applied Oral Science*, e20180693.

Thomas, B., Chandak, M. and Deosarkar, B. (2017). Comparison of acidic versus alkaline environment for furcation perforation repair among calcium silicate based materials: An *in vitro* comparative study. *British Journal of Medicine & Medical Research*, 1-8.

Thota, M. M., Kakollu, S., Chinthamreddy, S., Tammineedi, S., Bhupathi, Y. and Ediga, S. R. (2019). Comparative evaluation of sealing ability of ProRoot MTA, Biodentine, and Bone Cement in the repair of furcation perforation – An *in vitro* study. *Indian Journal of Dental Advancements*, 176-180.

Toia, C. C., Teixeira, F. B., Cucco, C., Valera, M. C. and Cavalcanti, B. N. (2022). Volumetric evaluation of voids and gaps of gifferent calcium-silicate based materials used in furcal perforations: A Micro-CT study. *Dentistry Journal (Basel)*, 41.

Torabinejad, M., Higa, R. K., McKendry, D. J. and Pitt Ford, T. R. (1994). Dye leakage of four root end filling materials: effects of blood contamination. *Journal of Endodontics*, 159-163.

Torabinejad, M. and Parirokh, M. (2010). Mineral trioxide aggregate: A comprehensive literature review – part II: leakage and biocompatibility investigations. *Journal of Endodontics*, 190-202.

Torabinejad, M., Parirokh, M. and Dummer, P. M. N. (2018). Mineral trioxide aggregate and other bioactive endodontic cements: an updated overview – part II: other clinical applications and complications. *International Endodontic Journal*, 284-317.

Torabinejad, M., Watson, T. F. and Pitt Ford, T. R. (1993). Sealing ability of a mineral trioxide aggregate when used as a root end filling material. *Journal of Endodontics*, 591-595.

Tsanova-Tosheva, D. M., Dimitrova, I. V. and Kouzmanova, Y. I. (2021). Analysis of the organic content in two hybrid calcium-silicate cements. In BECB 2021. *ACM'S International Conference Proceeding Series*, New York, USA, Article 12, 1–5.

Tsatsas, D. V., Meliou, H. A. and Kerezoudis, N. P. (2005). Sealing effectiveness of materials used in furcation perforation *in vitro*. *International Dental Journal*, 133-141.

Valois, C. R. and Costa, E. D. Jr. (2004). Influence of the thickness of mineral trioxide aggregate on sealing ability of root-end fillings *in vitro*. *Oral Surgery, Oral Medicine, Oral pathology, Oral Radiology, and Endodontics*, 108-111.

Von Loetzen, S. C. S. and Hulsmann, M. (2018). Root perforation repair concepts and materials: A review. *ENDO - Endodontic Practice Today (Lond Engl)*, 87-100.

Xavier, C. B., Weismann, R., de Oliveira, M. G., Demarco, F. F. and Pozza, D. H. (2005). Root-end filling: apical microleakage and marginal adaptation. *Journal of Endodontics*, 539-542.

Zafar, K., Jamal, S. and Ghafoor, R. (2020). Bio-active cements-Mineral Trioxide Aggregate based calcium silicate materials: a narrative review. *The Journal of the Pakistan Dental Association*, 497-504.

Biographical Sketches

Ivanka Veselinova Dimitrova

Affiliation:

Department of Conservative Dentistry, Faculty of Dental Medicine, Sofia Medical University.

Education:

Higher education (DMD), PhD degree, additional specialization in Health Management.

Research and Professional Experience:

Enamel Remineralization, Vital Bleaching, Bioactive Materials, Nano-technology.

1994 - defended thesis on topic: "Enamel remineral-ization – laboratory and clinical research."

Professional Appointments:

Associate Professor at Sofia Medical University,

Lecturer of Preclinical and Clinical Conservative Dentistry courses.

Publications from the Last 3 Years:

2020

Kouzmanova, Y. and I. Dimitrova. Solubility of calcium silicate-based cements. *Acta Medica Bulgarica.* 2020, Vol. XLVII, №2: 27-29. https://doi.org/10.2478/ AMB-2020-0019.

2021

Dimitrova I., Angelova L. and Hristov K. Study of frequency of iatrogenic perforations by group of teeth and localization made by Bulgarian students. *International J. of Current Research and Review.* IJCRR, 2021, 13, 21, 73-77. DOI: http://dx.doi.org/10.31782/IJCRR.2021.132113.

Dimitrova, I. and V. Dimitrova. Clinical case of conservative retreatment of compromised tooth. *Adv Dent & Oral Health,* 2021, 14(5): ADOH.MS.ID.555 900, 001-005.

2022

Dimitrova I.V. and Y. Kouzmanova. Evaluation of the levels of professional satisfaction among Bulgarian general dental practitioners: A pilot study. *Int. J. Adv. Res.* (IJAR). 2022 Apr; 10(4): 993-997.

Kouzmanova Y.I., I.V. Dimitrova and D.M. Tsanova-Tosheva. Long-term clinical repair of endodontic perforations treated with calcium silicate-based cements. 2021. In *2021 International Symposium on Biomedical Engineering and Computational Biology* (BECB 2021). Association for Computing Machinery, New York, NY, USA, Article 13, 1-3. doi: https://doi.org/10.1145/3502060.35 02319. Online: 14 February 2022.

Tsanova-Tosheva D.M., I.V. Dimitrova and Y.I. Kouzmanova. 2021. Analysis of the organic content in two hybrid calcium-silicate cements. In *2021 International Symposium on Biomedical Engineering and Computational Biology* (BECB 2021). Association for Computing Machinery, New York, NY, USA, Article 12, 1-6. DOI: https://doi.org/10.1145/3502060.3502276. Online: 14 February 2022.

Yaneta Ivanova Kouzmanova

Affiliation:

Department of Conservative Dentistry, Faculty of Dental Medicine, Sofia Medical University.

Education:

1980-1985 - Undergraduate medical education at Faculty of Dental Medicine, Sofia Medical University

1985 – Graduated with master's degree in Dental Medicine (DMD)

1990-1993 - Postgraduate medical education at Faculty of Dental Medicine

1993 – Acquired specialty: General Dental Medicine

1993 – present - Continuous postgraduate education in many courses in the field of endodontics.

Research and Professional Experience:

Endodontics, MTA and Calcium Silicate-Based Cements, Endodontic Anthropology, Endodontic Anatomy Aberrations, C-shaped Root Canal System, Endodontic Implantation (in the past).

2019 - defended thesis on topic: "Endodontic perforations – hermeticity of sealing with calcium-silicate cements, and prevention."

Acquired degree of Doctor in Medicine and Health Sciences (PhD).

Professional Appointments:

1985-1993 – Worked as a General Dental Practitioner at Worker's Hospital, Shumen

1993 – present – Works as a lecturer at Faculty of Dental Medicine, Sofia Medical University

2002 – present – Chief Assistant Professor at Faculty of Dental Medicine

Honors:

First award and Supportive award for Poster Presentations at the 5th and 6th International Congresses of Medical Sciences (ICMS) for medical students and young doctors, Sofia, Bulgaria

Publications from the Last 3 Years:

2020

Kouzmanova, Y. and I. Dimitrova. Solubility of calcium silicate-based cements. *Acta Medica Bulgarica.* 2020, Vol. XLVII, №2: 27-29. https://doi.org/10.2478/ AMB-2020-0019. Available from: https://www.researchgate.net/publication/3437209 83.

2022

Dimitrova, I. V. and Y. Kouzmanova. Evaluation of the levels of professional satisfaction among Bulgarian general dental practitioners: A pilot study. *Int. J. Adv. Res. (IJAR).* 2022 Apr; 10(4): 993-997. Available from: https://www.journal ijar.com/uploads/627cbcba17bbb_IJAR-39302.pdf.

Kouzmanova Y. Invasive cervical resorption due to iontophoretic internal bleaching with hydrogen peroxide: two case reports. *Journal of IMAB - Annual Proceeding (Scientific Papers).* 2022 Apr-Jun; 28(2):4361-4365. doi: https://doi. org/10.5272/jimab.2022282.4361. Available from: https://www.journal-imab-bg.org/issues-2022/issue2/2022vol 28-issue2-4361-4365.pdf.

Kouzmanova, Y. I., I. V. Dimitrova and D. M. Tsanova-Tosheva. Long-term clinical repair of endodontic perforations treated with calcium silicate-based cements. 2021. In *2021 International Symposium on Biomedical Engineering and Computational Biology (BECB 2021).* Association for Computing Machinery, New York, NY, USA, Article 13, 1-3. doi: https://doi.org/10.1145/3502060.3 502319.

Tsanova-Tosheva, D. M., I. V. Dimitrova and Y. I. Kouzmanova. 2021. Analysis of the organic content in two hybrid calcium-silicate cements. In *2021 International Symposium on Biomedical Engineering and Computational Biology (BECB 2021).* Association for Computing Machinery, New York, NY, USA, Article 12, 1-6. doi: https://doi.org/10.1145/3502060.3502276.

Chapter 2

Low-Energy Cements and Belite Activation

T. Staněk[1,*], PhD and P. Sulovský[2], PhD

[1]Research Institute for Building Materials, Brno, Czech Republic
[2]Faculty of Science, Palacky University, Olomouc, Czech Republic

Abstract

The term "low-energy cements" describes cements that can replace ordinary Portland cement in at least some applications and can also reduce energy consumption during the production process. Extension of these cements could lead both to lower costs of binders' production and lower emissions, CO_2 especially. Belite cements are part of low-energy cements. Pure belite clinkers with interstitial matter comprising C_3A and C_4AF are not produced, since they yield materials with insufficient strength. Currently, sulfoaluminate-belite cements, low-energy cements doped with fluorine and chlorine as well as undoped high-belite cements are produced in limited volumes in China. Results of hydraulic activation of high-belite clinker by sulfate anions are also given in this paper. The principle of activation is the preparation of belite clinker with increased Ca:Si ratio in the structure of dicalcium silicate by substitution a part of SiO_4^{4-} by SO_4^{2-}. Cements prepared from these high-belite clinkers with alite content up to 30% that were burned at 1,350°C display the same technological properties as ordinary alite Portland cements, including early strengths as ordinary alite Portland cements and comply with standard EN 197-1.

Keywords: Portland cement, low-energy cement, belite, hydraulic activation, sulfate

---

* Corresponding Author's Email: stanek@vustah.cz.

In: Properties and Uses of Calcium Silicate
Editor: Amanda G. Carlton
ISBN: 979-8-88697-128-6

Introduction

Portland cement is currently the most important and widely used hydraulic binder in the construction industry. The almost two centuries tradition of its production and use have proved its binding properties which are, together with a readily available material base and economically acceptable production costs, the main reasons for its production on such a large scale. Scientific projections indicate that despite the development and production of new types of binders, its importance will keep increasing (Uchikawa, 1994; Brandštetr, 1999).

The research of Portland cement also has a long tradition; however, there are still new areas emerging in both production and application and these areas are becoming subject of intense research. One of these areas is the search for effective ways of reducing energy consumption in the production of Portland cement. This issue is being currently very topical because the development of cement production is increasingly influenced by environmental and economic aspects and growing customers' demands for high quality of cement products.

As cement production is characterized by processing large amounts of natural materials and considerable energy consumption, producers' main aims include optimum utilization of primary raw materials, secondary raw materials, waste, secondary energy resources and achieving high quality and durability of cement products.

The annual global cement production reached 4.1 Mt in 2020; more than 58% was produced in China[1]. Cement production causes approximately 5% of global man-made CO_2 emissions[2].

The research and production of hydraulically active low-energy cements, especially of those based on high belite clinkers, is highly topical and for the future of cementitious binders highly prospective. It is expected to be one of the possible main directions of the development of world cement industry.

Mass production of low-energy cements would mean considerable decrease in total expenses on their production in comparison to the currently common Portland cement with high alite content, along with overall decrease

---

[1] The Statistics Portal, *Cement production globally and in the U.S. from 2010 to 2020*, (online), https://www.statista.com/statistics/219343/cement-production-worldwide/ (date of access: 2022-01-23).

[2] Fonta, P. Cement Sustainability Initiative - Cutting Global CO2 Emissions from Cement Production, *United States Climate Change* (online), https://unfccc.int/news/cement-sustainability-initiative-cutting-global-co2-emissions-from-cement-production (date of access: 2022-01-23).

of the environmental impact of cement production. The production of low-energy cements allows decreasing the temperature of burning by 100 – 300°C, what along with economic benefits brings also lowering of CO_2 and NO_x emissions. Further decrease of CO_2 emissions is brought by the composition of the raw meal for the production of low-energy clinkers, as they require less CaO and thus lesser consumption of $CaCO_3$. This approach also means saving resources of high-quality raw materials, above all pure limestones, and utilization of less valuable natural materials with lower $CaCO_3$ content, especially of various secondary and waste products from other industrial productions.

Belite in common Portland clinker has considerably lower hydraulic activity than alite (Sharara et al., 1994) and contributes significantly only to strengths after 28 days of hydration. This led to efforts to stabilize hydraulically active forms of belite, above all its high-temperature modifications. Chemical stabilization by suitable admixtures, usually complemented by extremely fast cooling (Chatterjee, 1996; Emanuelson, Landa-Cánovas and Hansen, 2003; Gies and Knöfel, 1986; Jost et al., 1988) is one of possibilities. A newer method of belite hydraulic activation is the utilization of the so called remelting reaction (Fukuda, Maki and Ito, 1992; Fukuda, Wakamatsu and Ito, 2001) or the sol-gel method (Chrysafi, Perraki and Kakali, 2007; Kurdowski, Duszak and Trybalska, 1997). Such methods can be realized only under conditions beyond the possibilities of currently used industrial technologies. The research of the mechanism and kinetics of belite clinker formation has shown, that quickly formed belite clinker has (in contrary to original expectations) lower hydraulic activity than longer burned, recrystallized belite clinker (Staněk, 2014).

Production of sulfoaluminate-belite (SAB) cements is performed in limited extent, albeit these cements show relatively good properties (Brown, 1992; Mojumdar and Janotka; 2002). Experiments with industrial production of sulfo-aluminate-ferrite belite clinker (SAFB) and high-Fe belite clinker (HFBC) yielded satisfactory strengths after 28 days of hydration, but low short-term strengths (Popescu, Muntean and Sharp, 2003).

China makes great progress in testing such technologies with industrial production of SAB and also fluoraluminate belite cements and belite-rich Portland cements with 20-30 wt. % of alite (Sui et al., 2005).

The main problem to be solved is the production of hydraulically active belite cement with properties like those of standard "alite" Portland cement, using the existing technological lines.

Portland Cement Clinker

The most important constituent of Portland cement is Portland clinker. The essence of its production is to transform a mixture of natural or artificial minerals into phases which spontaneously react with water to form new phases with a microstructure that works as a strength carrier. It lies basically in a change in the microstructure at the cost of supplied energy released as heat generated from cement hydration during setting. Therefore, cement producers aim to achieve such composition of phases that ensures the required properties of cement in the final product with minimum costs of energy for burning and grinding.

Raw Meal for Portland Cement Clinker Production

For all types of chemical production, including the production of Portland clinker, the most critical factor is the chemical composition of the initial reaction mixture – raw meal, consisting of particles of various chemistry, which, in a certain volume, constitute the chemical composition of clinker. The smaller and the more intermixed the particles are, the easier is clinker formed. This composition is ensured by proper proportioning of raw materials based on their chemical analysis. The basic oxides include:

CaO– limestone is the main source;
SiO_2– siliceous rocks, shales and clays are the main sources;
Al_2O_3–clays, clay constituents of limestones, shales and loams, and bauxite are common sources;
Fe_2O_3–industrial iron waste and iron ores are the most common sources.

The composition of raw meal from raw material constituents (minerals, industrial waste) is determined by its final chemical composition. Most often, the content of basic oxides in raw meal is expressed as their ratios expressed by the following formulae (i.e., chemical parameters):

Lime saturation factor after Lea – Parker: $SLP = \frac{100C}{2.8S+1.18A+0.65F}$ (1)

Lime saturation factor after Kühl: $SK = \frac{100C}{2.8S+1.1A+0.7F}$ (2)

Silica ratio: $SR = \frac{S}{A+F}$ (3)

Alumina-iron ratio: $AR = \frac{A}{F}$ (4)

where $C = CaO$, $S = SiO_2$, $A = Al_2O_3$, $F = Fe_2O_3$ in wt. %.

The right choice of chemical parameters for the chemical composition of raw meal is only a necessary but not sufficient condition for its high-quality preparation. Raw material constituents used for preparation of raw meal contain not only the basic oxides but have a significantly more complicated chemical composition. Besides the basic oxides, they also contain the so-called minor oxides, e.g., K_2O, Na_2O, SO_3, MgO, P_2O_5 etc., which may, at certain concentrations, significantly affect the process of clinker formation (Dreizler and Knöfel, 1982; Maki et al., 1991; Maki et al., 1992; Strunge, Knöfel and Dreizler, 1985, 1986; Osokin and Potapova, 1989; Maki and Goto, 1982; Ichikawa and Kanaya, 1997; Staněk and Sulovský, 2002). The basic criterion for the determination of suitability of a raw meal for clinker burning is, therefore, its overall chemical composition.

The second and equally important prerequisite is to put the entire system into such energy state in which the chemical balance is completely shifted towards the reaction products. As the desired clinker minerals are formed from raw meal only at very high temperatures (around 1,450°C), especially the rate of reactions forming clinker with the desired mineralogical composition is very important for the total energy performance of the process. The rate of clinker formation depends not only on the burning temperature but largely on the properties of raw meal, particularly on its reactivity. The second criterion of suitability of raw meal for clinker production is, therefore, its reactivity (Chromý, 1974a).

Grinding has a significant influence on the reactivity of raw meal because it increases with increasing meal fineness. Grinding fineness should be adjusted so that 10 - 30% material is retained on a 90 micron sieve (Hlaváč, 1988).

Phase Composition of Portland Cement Clinker

The main minerals of Portland cement clinker are silicates, namely alite (tricalcium silicate – Ca_3SiO_5 – C_3S) and belite (dicalcium silicate – Ca_2SiO_4 – C_2S). Both minerals are characterized by significant polymorphism.

Seven alite modifications (Bigare et al., 1967; Maki and Chromý, 1978a) differing structurally by minor deformations from the basic hexagonal structure, have been identified - three triclinic (T), three monoclinic (M) and one trigonal (R). Their transformation temperatures (°C) are shown below:

$$T_1 \overset{620}{\longleftrightarrow} T_2 \overset{920}{\longleftrightarrow} T_3 \overset{980}{\longleftrightarrow} M_1 \overset{990}{\longleftrightarrow} M_2 \overset{1060}{\longleftrightarrow} M_3 \overset{1070}{\longleftrightarrow} R$$

Alite in industrial clinker occurs only as M_1 and M_3 modifications, rarely T_2 modification (Maki and Chromý, 1978b). The proportion of alite nucleation and crystallization processes and the contents of secondary oxides in the melt are critical for stabilization of the above modifications (Maki and Goto, 1982; Staněk and Sulovský, 2002).

As for belite, five crystal modifications were discovered - α (hexagonal), α'_H (orthorhombic), α'_L (orthorhombic), β (monoclinic) and γ (orthorhombic) (Ghosh et al., 1979); occurrence of all these modifications in industrial clinker is possible. Regarding the stabilization of the individual modifications, the type and quantity of secondary oxides, burning temperature and cooling rate play important roles. Polymorphic transformations of belite occurring during cooling from temperatures above 1,400°C are shown below:

$$\alpha \overset{1425}{\longleftrightarrow} \alpha'_H \overset{1160}{\longleftrightarrow} \alpha'_L \overset{670}{\longleftrightarrow} \beta \overset{<500}{\longrightarrow} \gamma$$

β modification is metastable and in Portland clinker has to be stabilized using solid solutions to prevent its transformation to γ.

Two other basic phases create the clinker interstitial phases. The first one is dark coloured and contains ferric glass as well as imperfectly developed crystals which are similar by their composition to C_4AF mineral (brownmillerite). The light coloured connecting matter contains, in addition to glass with a low content of iron oxide, C_3A crystals or solid solutions thereof. C_4AF is referred to as the light interstitial phase and C_3A as the dark interstitial phase, based on the colour observed in reflected light in an optical microscope. Furthermore, clinker also contains a small amount of unreacted free CaO and free MgO (periclase), and alkali metal sulfates may also be present (Hlaváč, 1988).

Process of Portland Cement Clinker Formation

The creation of a liquid phase plays a significant role in the clinker formation process. Before its creation, at the start of the sintering zone in the furnace, raw meal decomposition processes (decomposition of carbonates, dehydration and dehydroxylation of clay minerals and micas) and solid phase reactions occur. These processes take place in current production systems in the dispersed state in heat exchangers (or, more precisely, in calciners), then, in a powdery state, in the rotary furnace, and altogether they are strongly endothermic. Belite is created by solid phase reactions during decomposition of limestone (Chromý, 1974a).

Clinker melt is formed at the beginning of the sintering zone. It is preceded by the formation of local areas of eutectics in particles with increased Fe content, which start the aggregation of loose material. The gradual formation of coherent clinker melt is initially accompanied by rapid exothermic belite formation (Chromý, 1982; Kondo and Choi, 1968; Christensen, Jepsen and Johansen, 1978) which promptly ends and the final phase of melt formation is already strongly endothermic.

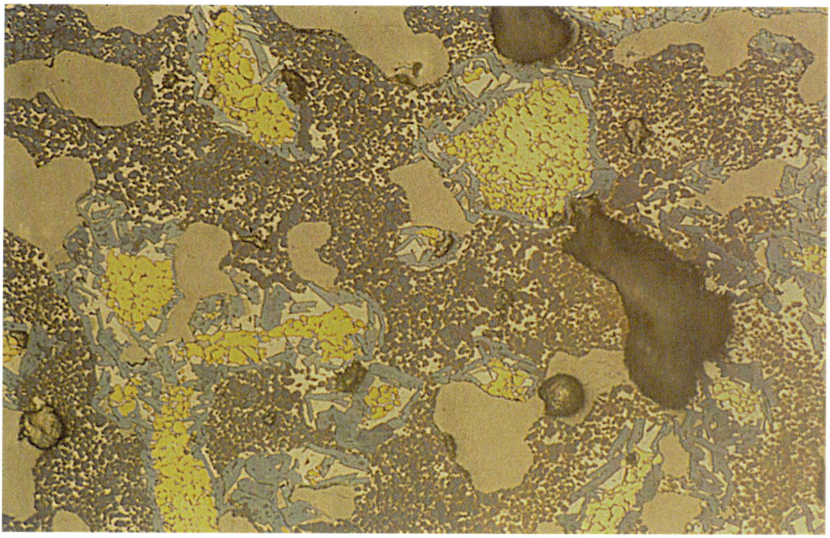

Figure 1. Microstructure of clinker at an early development stage. Belite (brown oval grains) prevails. Alite (blue tabular crystals) rims clusters of free CaO (yellow oval grains). Light matter between grains and crystals is solidified clinker melt; pores create beige (closed in epoxy resin) or black diffuse areas (open). Polished section, reflected light, dyed in acetic acid fumes, magnification 150 x.

Before the completion of belite formation (C_2S) through $2C + S \rightarrow C_2S$ reaction, a thin alitic layer, the so-called "limiting layer", is created around loose CaO clusters (see Figure 1). It is followed by the final, slowest, and therefore, crucial alite (C_3S) forming reaction– $C + C_2S \rightarrow C_3S$. Free CaO on the cluster surface dissolves in the melt, and is diffusionally transported through the liquid phase of the alite layer. Alite crystallizes on the outer surface of this layer immediately after the dissolution of belite. Since the difference of CaO concentrations in the melt between the surface of the free CaO cluster and the boundary between C_3S and C_2S is constant and very small (Johansen, 1979), the rate of this reaction is low and decreases progressively in with decreasing conversion rate (Christensen, Jepsen and Johansen, 1978; Chromý and Hrabě, 1982). Removal of the last percent of free CaO requires twofold elongation of isothermal clinker burning, which is a significant finding for the energetic economy of production.

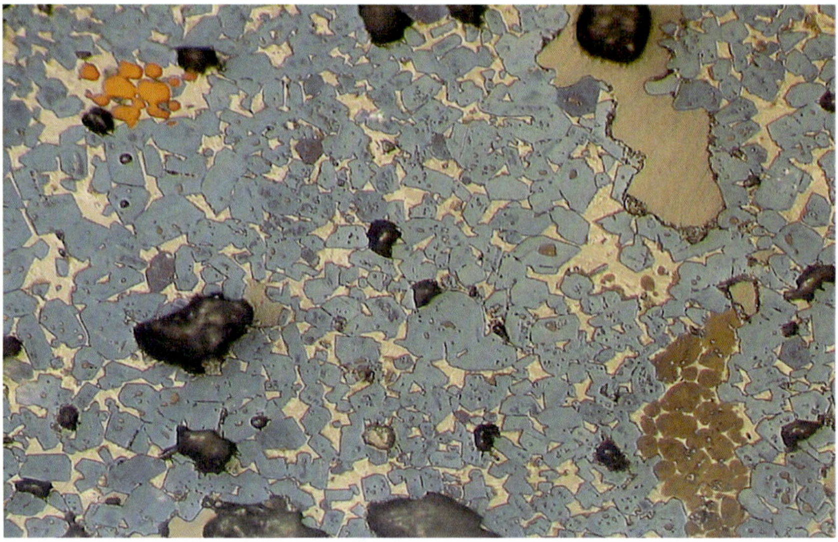

Figure 2. Microstructure of common Portland clinker with a high content of alite (blue crystals), low content of belite (brown grains) and residual free CaO (orange grains). Grey C_3A and white C_4AF can be seen in the clinker interstitial matter. Polished section, reflected light, dyed in acetic acid fumes, magnification 150x.

The creation of melt at the beginning of the sintering zone causes aggregation and granulation of the originally loose material. The intensity of granulation decreases with the increasing conversion of $C + C_2S \rightarrow C_3S$ reaction which depends on the increase of the content of alite in clinker.

Nodules with the majority of belite are plastic and their surface is intensively covered with melt (tighter packing of spherical belite grains which do not merge), while later the merging angular alite crystals (see Figure 2) create a firm frame of the nodule, less coated by the melt. Further aggregation of particles runs mainly by sticking of clinker dust on the nodules or by agglomeration of smaller granules. (Weber, 1989).

Low-Energy Cements

Sulfobelite Cement

Sulfobelite cements, also called sulfoalumiate-belite cements, contain C_2S and $C_4A_3\overline{S}$ (tetracalcium trialuminate sulfate, Klein's complex, yeelimite) as the main components. They do not contain alite and C_3A, however, they may contain varying amounts of calcium aluminoferrite. CA, $C_{12}A_7$, M (periclase), C_2AS (gehlenite), $C_5S_2\overline{S}$ (sulfospurrite) may also be present as minority phases (Andac and Glasser, 1995). C_3MS_2 (merwinite) may form in clinker prepared from materials with a high content of MgO (Ali, Sharma and Bhargava, 1992).

SAB cements solve the initial low hydraulic activity of belite by the presence of reactive phase $C_4A_3\overline{S}$, which ensures rapid growth of the initial strengths (Quillin, 2001). With sulfates, two ternary chemical compounds – $C_4A_3\overline{S}$ and $C_5S_2\overline{S}$ may form in the $CaO – SiO_2 – Al_2O_3 – SO_3$ system (Sahu, 1994).

Calcium sulfoaluminate can be created from the initial mixture with appropriate oxide composition by burning at 1,250 – 1,300°C (Santoro, Garofano and Valenti, 1986) according to the following reaction:

$$3CaCO_3 + 3Al_2O_3 + CaSO_4.2H_2O \rightarrow 4CaO.\ 3Al_2O_3.SO_3 + 3CO_2 + 2H_2O \qquad (5)$$

This compound is stable at temperatures below 1,350 – 1,400°C (Puertas, Varela and Molina, 1995). Sulfospurrite formed under common clinker production practice is stable only within a narrow range – 1,100 to 1,180°C – and is usually considered to have no important hydraulic properties, while

$C_4A_3\overline{S}$ creates ettringite when hydrated with calcium sulfate and lime (Hewlett, 2007).

The order of reactions with increasing temperatures in this system is described below:

- 800 – 900°C: creation of C_2S and C_2AS (gehlenite)
- Above 1,000°C: disappearance of C_2AS and creation of $C_4A_3\overline{S}$ and $C_5S_2\overline{S}$
- Above 1,180°C: disappearance of $C_5S_2\overline{S}$ and creation of α'-C_2S and $C\overline{S}$ (anhydrite)

Nevertheless, if sulfospurrite (ternesite) after its formation in the above temperature range is cooled gradually and relatively slowly, it does not decompose and acquires sufficient hydraulic reactivity (Bullerjahn et al., 2012).

Odler (2000) says that the terminology of various sulfobelite cements is rather confused and suggests the following classification of these cements:

- *Sulfoaluminate-belite cement (SAB)*: cement in which belite is the main constituent and the content of $C_4A_3\overline{S}$ is relatively low,
- *Sulfoferroaluminate-belite cement (SFAB)*: cement similar in composition to SAB but with a significant replacement of aluminium by iron,
- *Belite-sulfoaluminate cement (BSA)*: cement in which the sulfoaluminate phase predominates over belite,
- *Belite-sulfoferroaluminate cement (BSFA)*: BSA cement with a partial replacement of aluminium by iron in the sulfoaluminate phase,
- *Belite-sulfoaluminate-ferrite cement (BSAF)*: sulfobelite cement with a particularly high content of iron that – in addition to the Fe-doped sulfoaluminate – also contains an individual ferrite phase.

Examples of the structure of some sulfobelite cements and the development of their strengths are shown in Table 1.

Table 1. Effect of the composition of sulfobelite cements on the development of compressive strengths (Mehta, 1980)

Kind of cement	Fast hardening	Fast hardening	Normal	Slow hardening	Slow hardening
Phase, oxide	*Composition* [wt. %]				
C_2S	25	30	45	50	65
$C_4A_3\overline{S}$	20	20	20	10	10
C_4AF	40	30	15	30	15
Content of SO_3	15	20	20	10	10
Time	*Compressive strength* [MPa]				
8 hours	-	15.6	-	-	-
1 day	34.8	28.3	9.5	5.6	5.2
3 days	36.9	33.8	19.3	7.6	8.9
7 days	37.4	37.7	27.4	11.7	12.4
28 days	-	-	49.8	14.1	14.5
90 days	-	-	-	21.4	22.4
120 days	51.8	53.8	86.2	-	-

Sulfoalite Cement

Sulfoalite cement is characterized by the simultaneous presence of the phases C_3S and $C_4A_3\overline{S}$ in its clinker. Other phases may include C_2S, $C_2(A,F)$ and C_3A. The possibilities of producing such clinker initially appeared uncertain, as the optimum temperatures for synthesis of the two phases differ significantly: while $C_4A_3\overline{S}$ is formed at around 1,250 – 1,300°C and decomposes above 1,300°C, a temperature of at least 1,350°C is required to synthesize C_3S at an acceptable rate (Bullerjahn, 2012).

Moreover, the high content of SO_3 in clinker limits or blocks alite formation (Strunge, Knöfel and Dreizler, 1985, 1986; Staněk, 2016). However, it has been found that clinkers containing both phases may be produced at burning temperatures of 1,230 – 1,300°C, if small amounts of fluorspar (CaF_2) are added to the raw mix (Abdul-Maula and Odler, 1992).

An advantage of sulfoalite cements compared to sulfobelite cements is their more favorable strength development, especially between 1 and 28 days. Its grindability is also better (Abdul-Maula and Odler, 1992).

Sulfoaluminate Cement

Beretka et al. (1997) produced a series of cements that contained only the phases $C_4A_3\overline{S}$ (15 – 50%), $C_5S_2\overline{S}$ (sulfospurrite) (25 – 77%) and $C\overline{S}$ (anhydrite) (8 – 25%), but no calcium silicates. Such cements exhibit a very fast initial ettringite formation and strength development, but no additional strength gain at later ages.

Belite-Aluminate Cement

Aluminate-belite or belite-aluminate cements, also called porsal cements, contain β-C_2S calcium of CA and $C_{12}A_7$ aluminate phase as their main constituents. They can be produced by burning a raw material mixture consisting of limestone, clay and bauxite at temperatures below 1,250 – 1,300°C (Zakharov, 1974). Reaction products must be created by solid-state reactions because undesirable hydraulically inactive gehlenite is created during sintering in the presence of melt at higher temperatures.

Aluminate-belite cement contains 40 – 60% belite as the main constituent. If loss in strengths occurs due to the conversion of hydrated calcium aluminates, it is compensated by the strengths of hydrated components formed from β-C_2S (Zakharov, 1974; Viswanathan, Raina and Chatterjee, 1978). It is likely that this interception of the decrease of strengths is caused by formation of C_2ASH_8 (strätlingite) in the matrix, because it has been proved that C_2ASH_8 significantly contributes to compressive strengths of this matrix (Midgley and Bhaskara Rao, 1978).

Alite-Fluoraluminate Cement

Alite-fluoraluminate cement, also called regulated set cement (in the USA) or jet-cement (in Japan), contains C_3S (50 – 75%) and calcium fluoraluminate $C_{11}A_7.CF_2$ (20 –30%) as its main constituents. Other constituents may include the ferrite phase ($C_2(A, F)$), C_2S and free CaO (Odler, 2000).

Alite-fluoraluminate clinker can be produced from raw meals containing – in addition to CaO and SiO_2 – appropriate amounts of alumina and fluorspar (CaF_2), by burning them at temperatures of about 1,300°C (Chvatal, 1972; Odler and Abdul-Maula, 1980). The added CaF_2 acts as a flux, and allows the formation of C_3S at temperatures lower than usual in the production of ordinary Portland cement clinker.

Upon mixing with water, alite-fluoraluminate cement exhibits very rapid hydration. In paste hydration at ambient temperature about 40 – 50% of the $C_{11}A_7.CF_2$ is hydrated within the first 2 hours, and about 70% within 7 days (Uchikawa and Tsukiyma (1973); Knöfel D. and Wang, 1994). The hydration of C_3S is significantly slower: only about 10% of it hydrates within 2 hours.

Belite-Fluoraluminate Cement

These cements are made from clinker which contains $C_{11}A_7.CF_2$ and belite as the main phases but does not contain any alite. Clinker can be produced by burning a mixture of limestone, bauxite and fluorite in appropriate proportions in a rotary kiln or shaft furnace at temperatures ranging from 1,250 to 1,400°C (Liu et al., 1992).

The typical content of $C_{11}A_7.CF_2$ in these cements ranges from 40 to 80% and belite ranges from 10 to 55%. A limited amount of the ferrite phase may be present (Wang and Knöfel, 1994).

Ettringite is created as the main phase with aluminium hydroxide at the initial stage of hydration of these cements, while the C-S-H phase with portlandite is formed at the later hydration stages (Odler, 2000).

These cements may reach compressive strengths over 20 MPa after 2 hours of hydration and up to 60 MPa after 28 days (Wang and Knöfel, 1994) and they are intended for quick repairs in case of emergency when extremely fast hardening and increase of initial strengths is required.

Iron-Rich Belite-Fluoraluminate Cement

Clinkers used for preparation of these cements are characterized by the presence of a considerable proportion of the ferrite phase, in addition to calcium fluoraluminate and belite. Their composition may vary within the following range: $C_{11}A_7.CaF_2$ 5 – 60%, C_2S 10 – 50%, $C_2(A,F)$ 20 – 80% (Huang et al., 1992a).

Clinker can be burnt from a raw material mixture containing limestone, bauxite, iron ores and fluorite at temperatures ranging from 950 to 1,050°C. The ferrite phase in the form of C_6AF_2 is present in clinker burnt at such a low temperature, regardless of the A/F proportion in the initial raw meal (Huang et al., 1992b).

Compared to low-ferrous belite-fluoraluminate cements, the setting time of these cements is considerably longer, however, it does not exceed 1 hour (Huang et al., 1992b; Ikebata, Miyama and Sato, 1994).

Belite Fluor-Sulfo-Aluminate Cement

Clinker for this cement contains both $C_{11}A_7.CF_2$ and $C_4A_3\overline{S}$ together with belite and anhydrite. It may also contain a limited amount of the ferrite phase (Odler, 2000).

The highly ferrous belite fluor-sulfo-aluminate cement is characterized by a high content of Fe and the presence of a considerable amount of the ferrite phase in clinker. The composition of clinker may vary within the following range: $C_2(A, F)$ 30 – 75%, C_2S 0 – 23%, $C_{11}A_7.CF_2$ 0 – 32%, $C_4A_3\overline{S}$ 17 – 34% (Feng and Zhu, 1986).

Alumina-Ferrite and Belite Sulfo-Ferrite Cement

Ferrous bauxite, limestone and gypsum can be used for production of ferroaluminate cement clinkers in CaO-SiO_2-Al_2O_3-Fe_2O_3-SO_3 system by burning at 1,350 ± 50°C. These products are part of a group of special cements which require relatively low energy consumption, lower sintering temperatures and contain a low proportion of CaO in the basic phases, such as belite and calcium aluminoferrite. Their phase composition is: $C_4A_3\overline{S}$ 35 – 60%, C_4AF 15 – 45% and C_2S 15 – 30%. Consumption of energy necessary for their production is 65% lower than for the production of common Portland cement. These cements are highly resistant to sulfates and reach compressive strengths around 20 MPa after 1 day and around 100 MPa after 3 days (Wang and Deng, 1986).

Lattice defects caused by the presence of impurities and the low burning temperature contribute to their enhanced hydraulic properties (Boikova, Fomicheva and Grishchenko, 1992).

In CaO-F_2O_3-$CaSO_4$ system sintering starts at 800°C with the formation of CF. At temperatures ranging from 950 to 1,205°C, CF and $CaSO_4$ react to calcium monosulfoferrite $C_4F_3\overline{S}$ which decomposes at 1,205°C and $C_3F\overline{S}$ is created. Clinkers containing sulfoferrite may be burnt at 1,200 – 1,350°C; they are highly resistant to salt water effects and their compressive strength exceeds 80 MPa after 28 days of hydration. Hydration of calcium sulfoferrites and addition of sulfoferrite clinker to Portland cement improves its properties (Osokin, Krivoborodov and Dyukova, 1992).

Alinite Cement

Alinite cement was developed in the former Soviet Union, and was patented by Noudelman in 1977. It is characterized by the presence of the alinite phase in clinker instead of alite.

Alinite is a calcium oxy-chloro-aluminosilicate ($Ca_{11}[3(SiO_4).(AlO_4)]O_2Cl$ (Ilyukhin et al., 1977), which is related to alite, the main constituent of Portland cement. The main feature of alinite is the peculiar position of chlorine atoms surrounded by eight calcium atoms (Noudelman and Gadaev, 1986). Alinite can be synthesized from a raw meal containing $CaCO_3$, SiO_2, Al_2O_3, MgO and $CaCl_2$ by burning it at a temperature of 1,000 – 1,200°C (Bikbaou, 1986). Alinite exhibits a high hydraulic reactivity, which is distinctly greater than that of tricalcium silicate (Boikova, Grishchenko and Domansky, 1986).

Alinite clinker may be produced from a raw meal containing limestone, clay, MgO and $CaCl_2$ (6 – 18 wt. %) by burning it to temperatures of 1,000 – 1,300°C. A variety of industrial waste products, such as fly ash, magnesite dust, steel plant wastes, or municipal wastes with a significant amount of chlorine-containing materials, may be also utilized as raw meal constituents (Bikbaou, 1986). The phase composition of alinite clinkers varies typically in the following range: 50 – 80% of alinite, 10 – 40% of belite, 5 – 10% of calcium aluminochloride and 2 – 10% of calcium aluminoferrite (Pradip and Kapur, 1990).

Alinite cement exhibits fast initial hydration, which exceeds that of Portland cement (Ji, Ren and Su, 1997). The initial strength development of alinite cement is similar or even superior to that of ordinary Portland cement, but the 28-day strength is lower (Pradip and Kapur, 1990).

The use of alinite cement in steel-reinforced constructions must be discouraged because of possible corrosion of the reinforcement.

Fluoralinite cement is a fluorine analogue of alinite cement, and is characterized by the presence of fluoralinite ($C_{19}S_7.CaF_2$) (Ludwig and Urbonas, 1993). Fluoralinite cement clinker may be produced by burning suitable raw meal with fluorspar (CaF_2) content between about 3.5 and 8 wt. %, at temperatures of 950 – 1,100°C. The clinker contains between 35 and 80 wt. % of the fluoralinite phase and it can contain also C_2S, C_3S, $C_{11}A_7.CaF_2$ and C_4AF. In contrast to alinite cement, fluoralinite cement does not cause corrosion of the reinforcement steel in the resulting concrete.

Belinite Cement

A phase similar to belite containing chlorine is named belinite and its chemical formula is $Ca_8Mg(SiO_4)_4Cl_2$. Belinite has been synthesized from $CaCl_2$ melt containing MgO (Lampe et al., 1986).

Compared to alinite, its hydraulic properties are quite low and comparable to β-dicalcium silicate (Lampe et al., 1986; Jost and Ziemer, 1984).

Roman Cement

Roman cement was a predecessor of the present Portland cement. It used to be made from calcinated lumps of clayey limestones at temperatures below the sintering level, approximately 900 – 1,000°C. The binder made as described above was improperly called "Roman cement", although it is not similar to binders used in mortars in ancient Rome (Hewlett, 2007).

The cement contained mainly C_2S, CA, C_4AF and free CaO, and did not contain C_3S. The composition indicates quick setting due to the CA hydration and rather low strengths. When the initial material contained a high proportion of MgO, it could have contained a high proportion of free MgO which might have resulted in volume changes occurring during setting (Grosvalds, Lagzdina and Sedmalis, 1997).

Roman cement was invented in England and patented in 1796. It used to be widely used in the first half of the 19th century. Approximately after 1850 it was gradually replaced by Portland cement (Hewlett, 2007), although continental European market for Roman cement was more prolonged, as witnessed by many late 19th century buildings, where it was used owing to the prolonged plasticity and workability in mortars for rendering and cast decorative ornament, characteristic for the secession (jugendstil, decorative style) architecture (Kozłowski, Hughes and Weber, 2010).

Production and Utilization of Low-Energy Cements

Production of sulfoaluminate-belite (SAB) cements is performed in limited extent, as these cements show appropriate properties (Brown, 1992; Mojumdar and Janotka, 2002). The main hydration product of these cements – ettringite – with an extreme content of crystal water ensures rapid growth of the initial strength, however, the doubts about long-term stability of concrete made from these cements remain.

Experiments with industrial production of sulfo-aluminate-ferrite belite clinker (SAFB) and high-Fe belite clinker (HFBC) were realized, which yielded satisfactory strengths after 28 days of hydration, but low short-term strengths (Popescu, Muntean and Sharp, 2003).

China makes great progress in testing such technologies with industrial production of SAB and also fluoraluminate belite cements and high-belite Portland cements with 20-30 wt. % alite (Sui et al., 2008).

Chinese authors describe 6 types of low-energy cements produced (Sui et al., 2008; Huang et al., 1992; Wang and Su, 1989):

- *System of CaO-SiO_2-Al_2O_3 (NA):* Content of CaO in the system is such that only minerals with low content of CaO such as CA, CA_2 and C_2S are formed. Clinker is burned at 1,400°C from raw mix of limestone, quartz and high quality Al_2O_3.
- *System of CaO-SiO_2-Al_2O_3-$CaCl_2$ (LC):* $CaCl_2$ considerably decreases the temperature of sintering. Clinker contains alinite that is stable from 1,050 to 1,250°C, phase $11CaO{\cdot}7Al_2O_3{\cdot}CaCl_2$ that is formed from 750°C and it is stable up to 1,300°C and C_2S. Sintering temperature is 1,300°C and clinker is produced from limestone, fly ash and alkali slag.
- *System of CaO-SiO_2-Al_2O_3-CaF_2 ($\overline{F}A$):* CaF_2 significantly decreases the sintering temperature. Clinker consists of $11CaO{\cdot}7Al_2O_3{\cdot}CaF_2$ and C_2S and can contain $3CaO{\cdot}SiO_2{\cdot}CaF_2$, C_3A and C_3S. The raw mix of limestone, bauxite and fluorite is used for clinker production. Sintering temperature is 1,300°C.
- *System of CaO-SiO_2-Al_2O_3-$CaSO_4$ ($\overline{S}A$):* Clinker consists mainly of $C_4A_3\overline{S}$ and C_2S, and can contain of $C_{12}A_7$, CA and $C\overline{S}$. Clinker is burnt from mixture of gibbsite, limestone and gypsum. Sintering temperature is 1,350°C.

- *System of CaO-SiO_2-Al_2O_3-Fe_2O_3-$CaSO_4$ (FA):* This clinker contains ferrite phase (solid solution C_2F – C_6A_2F) same as ferrite phase from Portland cement clinker. Furthermore, it contains $C_4A_3\overline{S}$, C_2S and $C\overline{S}$. Clinker is prepared from raw mix of limestone, Fe-Al slag and gypsum at sintering temperature of 1,300°C.
- *System of CaO-SiO_2-Al_2O_3-Fe_2O_3-$CaSO_4$-CaF_2 (HCA):* In this system, unlike in the system without fluorite, C_3S can be formed. Clinker may contain C_3S, $C_4A_3\overline{S}$, C_6A_2F and $C_{11}A_7{\cdot}CaF_2$ or C_3A. The raw mix contains limestone, clay, gypsum and fluorite (or slag). Sintering temperature is 1,300°C.

Table 2. Compressive strength for 6 types of low-energy quick-hardening cements in MPa (Wang and Su, 1989)

Kind of cement	Hydration period			
	12 hours	1 day	3 days	28 days
NA	31.3	43.6	51.5	80.0
$\overline{S}A$	37.3	67.2	81.8	-
FA	-	57.9	77.6	85.4
HCA	-	34.3	46.3	63.5
$\overline{F}A$	31.4	40.9	-	53.8
LC	-	38.6	54.5	83.4

The development of compressive strength for these 6 types of low-energy cements is shown in Table 2. $\overline{S}A$ cements may be used for winter concrete work down to -25°C. Quick-hardening FA cements are highly resistant to the sea-water attack and, therefore, they are used for quick repairs and structures in marine conditions. NA and $\overline{F}A$ cements may be used for repairs on airport runways. Slightly expansive $\overline{S}A$ and FA cements were used for repairs of damaged cellars. $\overline{S}A$ and FA cements may be applied for guniting. Self-expansive NA, $\overline{S}A$ and FA cements are used for production of pressure pipelines of various diameters.

Common Portland cements with a high content of belite (HBC) are also manufactured in China; they are used in massive and cast concretes and in concretes with high technical parameters. Compared to common Portland alite cement, HBCs show better parameters as workability, heat release, higher long-term strengths and durability. Table 3 shows comparison of HBC and Portland cement (Sui et al., 2004).

Table 3. Example of properties of high-belite cement (HBC) and common Portland cement (PC) (Sui et al., 2004)

Kind of cement		PC	HBC
Phase composition [wt. %]	C_3S	53.6	26.4
	C_2S	20.6	50.8
	C_3A	7.5	3.7
	C_4AF	10.6	15.1
Hydration heat [kJ/kg]	7 days	297	232
	28 days	357	270
	180 days	406	321
Flexural / compressive strength [MPa]	3 days	5.5 / 31.4	4.1 / 16.5
	7 days	7.1 / 42.3	4.8 / 23.4
	28 days	8.0 / 51.5	8.7 / 50.1
	90 days	8.8 / 65.6	10.2 / 79.6
	180 days	10.0 / 68.6	11.6 / 83.1
	365 days	10.1 / 71.0	12.0 / 86.2

The typical mineral composition of HBC clinker is 20 – 30% C_3S, 45 – 60% C_2S, 3 – 7% C_3A and 10 – 15% C_4AF. Production of this clinker shows the advantages of low burning temperatures – around 1,350°C, energy saving up to 20%, approximately 10% reduction of CO_2 emissions, use of poorer quality limestone and the total costs of its production are 20% lower compared to common Portland cement (Sui, Liu and Guo, 1998).

Dicalcium Silicate Doped with SO_3

Preparation and Method

To prepare pure dicalcium silicate doped with SO_3, chemically pure $CaCO_3$, $CaSO_4.2H_2O$ and quartz sand (with 99.88 wt. % SiO_2). The carbonate and sulfate used were fine precipitates; quartz sand was ground to <0.04 mm.

The mixing of compositions was made in two ways. The first way keeps the stoichiometric ratio of CaO to SiO_2 of C_2S also after the addition of $CaSO_4.2H_2O$, in the second way pure C_2S makes 100% and additions of both oxides, CaO and SO_3 are as a surplus (in designations of samples marked by + sign).

The following Tables 4 and 5 give composition of individual mixtures for the preparation of dicalcium silicates doped with SO_3, which were chosen for burning for detailed research. The proposed mixtures were exactly weighted and homogenized.

Table 4. Mixtures for the preparation of C_2S with graded SO_3 content (in wt. %)

Designation	$CaCO_3$	Quartz	$CaSO_4 . 2H_2O$
C_2S - 5S	70.67	22.42	6.91
C_2S - 8S	67.10	22.04	10.86
C_2S - 10S	64.78	21.79	13.43
C_2S - 15S	59.22	21.19	19.59

Table 5. Mixtures for the preparation of C_2S with graded SO_3 content and surplus CaO (in wt. %)

Designation	$CaCO_3$	Quartz	$CaSO_4 . 2H_2O$	CaO surplus
C_2S - 5S+	69.45	20.84	9.71	3.5
C_2S - 8S+	65.63	10.70	14.68	5.6
C_2S - 10S+	63.30	19.00	17.70	7.0
C_2S - 12S+	61.14	18.35	20.51	8.4
C_2S - 15S+	58.16	17.46	24.39	10.5
C_2S - 18S+	55.45	16.65	27.90	12.6
C_2S - 20S+	53.79	16.14	30.07	14.0

Mixtures given in Tables 4 and 5 were burned in a programmable superkantal oven at temperatures of 1250 and 1450°C for 120 minutes.

The mentioned burnings were preceded by preliminary experiments, which led to the design of these compositions. A pure C_2S burned at 1500°C, has been immediately after cooling disintegrated into a dust composed of γ-C_2S. It is known that β-C_2S modification is metastable and can be stabilized against transformation to γ-C_2S by solid solutions in the crystal lattice. Furthermore, burnings of C_2S from the initial mixture with the addition of 0.5, 1, 3, 5, and 7% SO_3 at the beginning of burning were performed. In common clinker in the presence of alkalis and at addition of SO_3 alkali sulfates form preferentially. In the absence of alkalis, SO_3 enters the structure of belite (Gies and Knöfel, 1987) and, depending on the temperature of burning, partially volatilizes. These burnings have shown that with increasing SO_3 content in the starting mixture the intensity of the transformation into γ-C_2S decreases along with the decrease in the content of α'-C_2S modification, and the β-modification becomes stabilized and its content grows.

Furthermore, sandwich experiments have been realized with the aim of following the entry of SO_3 into the C_2S structure directly on the contact of C_2S with the bearer of SO_3. Anhydrite ($CaSO_4$) was used as the phase providing SO_3. Pressed two-layer pellets from C_2S and anhydrite were burned at 1450°C with different holding times (0, 30 and 60 minutes). Another experiment of this type was done with small pellets from C_2S or from anhydrite, which were enveloped by anhydrite, or C_2S and pressed into the form of bigger pellets, which were then burned at 1250 and 1350°C.

The phase composition of samples and proportions of individual belite properties were studied by X-ray diffraction, using the diffractometer D8 Advance (Bruker) with Cu anode ($\lambda K\alpha = 1.54184$ Å) in the angular range of 6-80° degrees 2 Θ.

The pellets of both types were cut perpendicularly to the C_2S/anhydrite interface(s), mounted in epoxy resin, polished and coated with carbon for electron microprobe analysis. The polished sections were analyzed with electron microprobe CAMECA SX100 with 5 crystal spectrometers to determine the chemical composition of clinker minerals. The quantitative spot analyses were performed under the following analytical conditions: accelerating voltage 15 kV, probe current 20 nA, spot size 2 microns. The acquired intensities were converted to concentrations using the PAP correction scheme. The used reference materials were natural minerals wollastonite (CaKα, SiKα), jadeite (NaKα), orthoclase (KKα, AlKα), anhydrite (SKα), hematite (FeKα), and synthetic periclase (MgKα) and TiO (TiKα).

Results

After the burning at 1450°C it was found that the sample C_2S - 5S is strongly disintegrated, the sample C_2S - 8S slightly and C_2S - 10S negligibly disintegrated; the sample C_2S - 15S was distinctly porous and sintered. All samples of the S+ series were solid, without signs of disintegration, slightly sintered. The burning at 1250°C yielded all samples stable, but less solidified (sintered) than those burned at 1450°C.

In the experiments with layered pellets SO_3 was strongly volatilized and its contents found in belite were low. The pellets with anhydrite core disintegrated after the burning, and γ-C_2S formed. The most promising results for further experiments were obtained with pellets consisting of C_2S core, enveloped by anhydrite.

Microscopic study has revealed that from raw meals with SO_3 addition the burning at 1450°C gave birth to a texture of closely packed belite grains with different degree of porosity. At 1250°C originates a porous fine grained multiphase aggregate, containing also a smaller amount of belite. In case of the sandwich pellets, the interface belite – anhydrite is sharp. The belite layer is formed by spherical grains cemented by congealed sulfate melt.

Figure 3. Microstructure of the mixture for preparation of sulfur-doped dicalcium silicate burned at 1250°C (backscattered electron image – BEI).

The examples of microstructure of burned samples photographed in SEM are given in Figures 3 and 4.

The X-ray diffraction analysis of samples C_2S - 5S to C_2S - 15S burned at 1250°C has shown that they contain β-C_2S, free CaO, secondary portlandite ($Ca(OH)_2$), anhydrite II ($CaSO_4$) and ternesite ($Ca_5(SiO_4)_2SO_4$). Also, the presence of wollastonite could not be excluded.

Figure 4. Microstructure of the mixture for preparation of sulfur-doped dicalcium silicate burned at 1450°C (BEI).

With increasing SO_3 content the proportions of anhydrite and ternesite increase and the amount of β-C_2S decreases. Ternesite is in the cement terminology called sulfate spurrite and occurs as a minor phase in sulfoaluminate-belite clinkers (Andac and Glasser, 1995; Sahu, 1994). Sulfate spurrite is stable in a narrow temperature range (1100 – 1180°C) and does not show significant hydraulic properties (Hewlett, 2007). The samples C_2S-5S+ to C_2S-20S+ burned at 1250°C had qualitatively the same phase composition; the difference was in the fact that sample C_2S-5S+ contained also a lesser amount of γ-C_2S and samples C_2S-8S+ to C_2S-12S+ contained also some tricalcium silicate (C_3S). The growing addition of SO_3 has in this series of samples increased the ternesite proportion and decreased that of β-C_2S. Comparing the samples with equivalent addition of SO_3 the first series has lower ternesite and higher β-C_2S and anhydrite than samples from the series with surplus CaO.

After the burning at 1450°C the samples C_2S - 5S to C_2S - 15S contain dicalcium silicate in α', β as well as γ modifications, C_3S, free CaO and secondary portlandite. The sample C_2S - 15S does not contain γ-C_2S or C_3S.

The increasing content of SO_3 decreases the amount of γ-C_2S and α'-C_2S, and increases β-C_2S. The samples C_2S-5S+ to C_2S-20S+, burned at 1450°C have similar phase composition, but they do not contain γ-C_2S. The phase composition in this series does not show any clear trend, as the bulk SO_3 content in these samples is roughly the same, amounting to 2.0 – 2.5 wt. %.

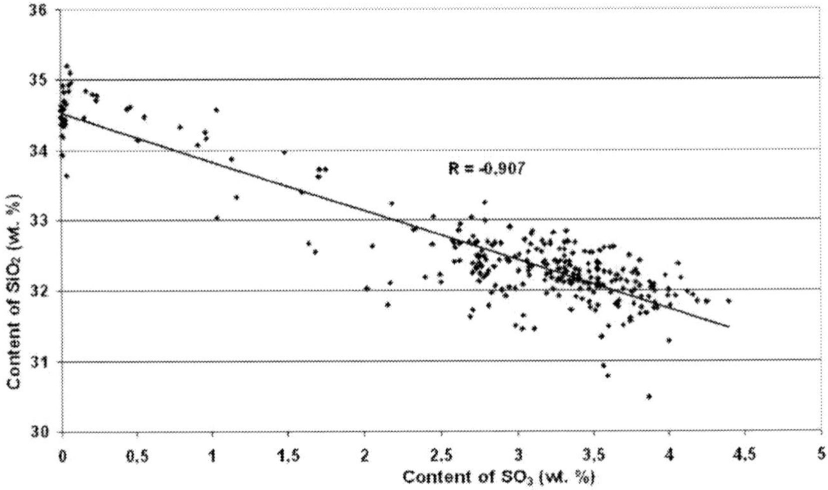

Figure 5. The relation of SiO_2 and SO_3 contents in belite with regression line and the correlation coefficient R.

The electron microprobe analysis of belite grains in experimental samples has revealed significant levels CaO, SiO_2 a SO_3 only. MgO was found on the 0.X wt. % level (max. only around 0.5 wt. %); the other oxides were present in amounts of the 0.0X wt. % order. Based on the burning temperature, all results were summarized into two datasets (burned at 1250°C / burned at 1450°C + sandwiches). Statistical analysis of the data focused on the dependence of CaO, SiO_2 contents and CaO : SiO_2 ratio on the content of SO_3. Samples burned at 1250°C did not show any correlation among these parameters. Samples burned at 1450°C, including the samples from sandwich experiments, have shown close negative correlation (R = -0.907) between SiO_2 and SO_3 contents in belite (see Figure 5). At the same time, SiO_2 negatively correlates with CaO (R = -0.776) and positive correlation exists between SiO_2 and the CaO : SiO_2 ratio (R = 0.783) – see Figure 6.

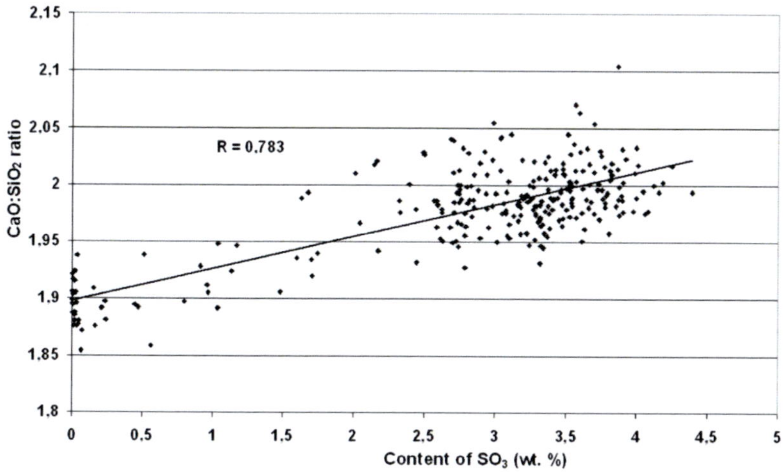

Figure 6. The dependence of the CaO:SiO$_2$ ratio on the SO$_3$ content in belite with regression line and the coefficient of correlation R.

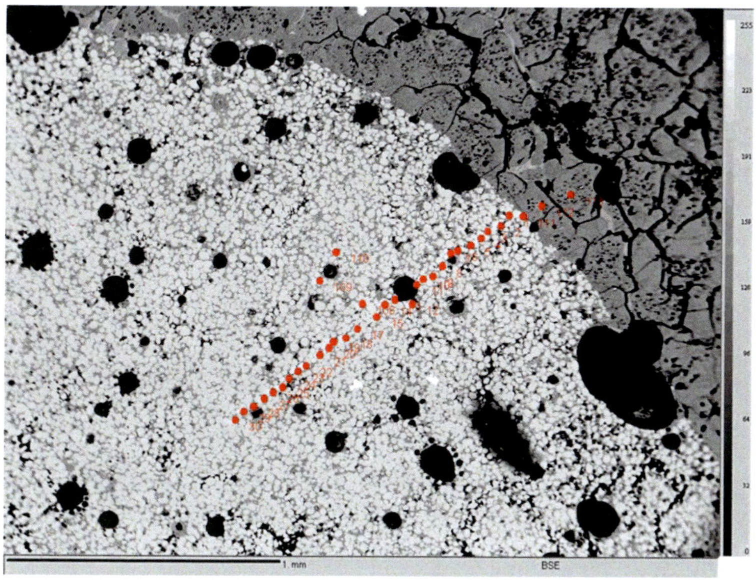

Figure 7. Sandwich microstructure with marked analysis points: belite (left) – anhydrite (right above) (BEI).

Microanalysis of belite grains at the contact with anhydrite in sandwich samples did not reveal any change in their chemistry up to the distance of 1.25

mm from the contact (see Figures 7 and 8). The contents of SO_3 found in these experiments document the maximal possible entry of SO_3 into the dicalcium silicate structure under the common burning conditions.

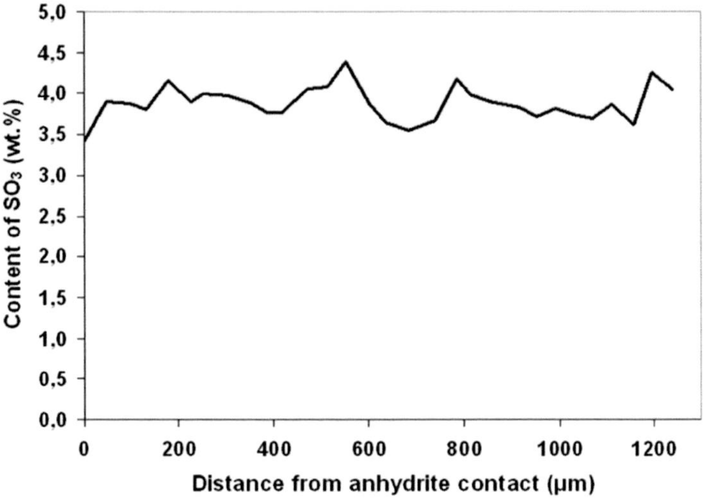

Figure 8. Content of SO_3 in belites at the contact with anhydrite (wt. %).

X-ray diffraction study has shown that belites with 2.4 – 3.2 wt. % SO_3 belong almost exclusively to the β modification.

Discussion

The research has shown that sulfur substitutes in the structure of dicalcium silicate silicon, what causes the stabilization of its β modification. This stabilization has already been mentioned in literature (Morsli et al., 2007). Sulfur enters the belite structure in the form of sulfate anion; although the ionic radius of SiO_4^{4-} is 40 pm and that of SO_4^{2-} only 29 pm, in spite of that it was observed that increasing incorporation of sulfate anion into calcium disilicate slightly enlarges its cell volume (Morsli et al., 2007).

Maximum content of SO_3 in dicalcium silicate was found by sandwich experiments; it amounts under common burning conditions to about 4.4 wt. %. The empirical formula of belites most rich in sulfur is:

$$Ca_{1.929\pm0.004}Mg_{0.014\pm0.002}\square_{0.055}[(SiO_4)_{0.902\pm0.002}(SO_4)_{0.082\pm0.002}]$$

where □ - vacancy, average of 30 analyses, 95% confidence intervals.

Taylor (Taylor, 1999) has found by analysis of industrial clinkers, that at bulk SO_3 content in clinker up to 3 wt. % the concentration of SO_3 in belite seldom exceeds 2 wt. %, bulk contents of SO_3 higher than 3 wt. % are rare in industrial clinkers.

The incorporation of SO_3 into the belite structure leads to the decrease of CaO content, but at the same time the Ca:Si ratio grows from about 2.03 ± 0.01 at 0 wt. % SO_3 in belite up to 2.16 ± 0.01 at 4.4 wt. % SO_3 in belite. The coefficient of correlation between the atomic Ca:Si ratio and the number of S atoms per formula unit (a.p.f.u.) is 0.79, what at the high number of analyses (>300) is statistically very significant value. The CaO contents decreases from about 65.6 wt. % to 63.7 wt. %.

Belite Clinkers Doped with SO_3

Preparation of Clinkers and Methods of Their Study

The raw materials used for the preparation of experimental raw meals were common cement-making materials – pure limestone (L1), limestone with increased SiO_2 content (L2), clay shale (CS), iron correction (Fe) and FGD gypsum (EG). The composition and expected chemical parameters of the raw meals are given in Table 6. The planned experiments involved burning of two belite clinkers with about 4 wt. % SO_3 and different lime saturation index, the reference low-saturation belite clinker, and reference alite clinker without SO_3 addition.

6 kilograms of each raw meal were ground in a laboratory ball mill to fineness characterized by - 0.09 mm fraction of about 12 wt. %.

The results of optical microscopy of clinkers prepared by preliminary equilibrium burnings were used to correct the composition of all raw meals by addition of such SiO_2 amounts, that the clinkers had the calculated phase composition.

The raw meals with corrected composition were pressed into tablets 4 cm in diameter, weighing about 80 g. All belite clinkers were burned in a superkantal oven in the following regime: rate of temperature rise 15°C/min, final temperature 1400°C, isothermal dwell 40 minutes. The alite clinker was burned at 1450°C for 120 minutes.

Table 6. The composition of raw meals in wt.% and expected basic chemical parameters

Sign	RM-B	RM-B80S4	RM-B90S4	RM-A
L1	55.27	48.82	52.74	64.41
L2	19.59	18.92	17.51	14.1
CS	23.72	22.96	20.57	20.62
Fe	1.42	1.37	1.26	0.87
EG	-	7.92	7.92	-
LSF	80	80	90	96
SR	2.6	2.6	2.6	2.6
AR	1.4	1.4	1.4	1.6

The representative portion of clinker needed for the determination of phase composition was crushed and sieved to obtain 0.045 – 1 mm grain size fraction, which was then mounted in epoxy resin and polished for observation in optical microscope. To enable the perfect distinction of clinker phases, the polished section was etched in acetic acid fumes (Chromý, 1974b).

Thus, prepared polished and etched sections were point counted. To obtain accurate volume proportions of clinker phases, 2000 points were registered (Chromý, 1978). For the recalculation from volume to weight percentages, the following densities were used: C_3S – 3.15, C_2S – 3.28, C_3A – 3.03, C_4AF – 3.77, free CaO – 3.35 $g.cm^{-3}$.

After repolishing, the sections were analyzed with electron microprobe (see above). The qualitative composition of SO_3-containing belite clinkers was also determined by X-ray diffraction.

Subsequently, cements with approximately the same specific surface (ca 440 m^2/kg) were prepared from all clinkers, to which 4 wt. % of natural gypsum were added as controller of setting. The cements were subjected to technological tests according to the European standards of the EN 196 series.

The determination of the released hydration heat was done by the dissolution method according to EN 196-8 after 2, 7, 28, 56, and 90 days of cement paste hydration.

The temperature changes during the cement paste hydration were followed in semiadiabatic calorimeter (Brandštetr et al., 2001), complying with the EN 196-9 standard. The calorimetric curve gives information on the intensity of hydration and the rate of hydration of the cement or binder. The calorimetric curves were also used to calculate the specific hydration heat of cements during the 24 hours of hydration according to the following formula:

$$Q_{24} = \frac{C_{total} \times (T - T_0)}{m} \quad (6)$$

where:

C_{total} – total thermal capacity of the calorimeter in J.K^{-1} (cement, water, vessel)
m – weight of cement in g
T, T_0 – ending and starting temperature of the mixture

Results

Composition of Clinker

The results of quantitative phase composition of all clinkers are given in Table 7 and the chemical parameters in Table 8. Figures 9 and 10 demonstrate the microstructure of clinkers containing SO_3.

Table 7. Phase composition of clinkers in wt.%

Clinker	S-B	S-B80S4	S-B90S4	S-A
C_3S	0.2	0.0	8.5	66.7
C_2S	79.3	82.6	73.3	12.2
C_3A	4.0	2.6	4.2	12.3
C_4AF	16.5	14.8	14.0	7.8
C_{free}	0.0	0.0	0.0	0.9
C_3S_{equil}	0.2	0.0	8.5	70.5
C_2S_{equil}	79.3	82.6	73.3	9.4

where:

$C_3S_{equil} = 4.219 * C_{free} + C_3S$ calculated equilibrium content of alite (7)

$C_2S_{equil} = 100 - C_3S_{equil} - C_3A - C_4AF$ calculated equilibrium content of belite (8)

Note that the phase list does not involve phases not recognizable in reflected light (anhydrite II, according to Rietveld analysis around 2 wt. %) or present in minute quantities (yeelimite); the percentages of phases identified are recalculated to total of 100%.

Table 8. Chemical composition of clinkers in wt. % and thereof calculated chemical parameters

Clinker	S-B	S-B80S4	S-B90S4	S-A
SiO_2	27.03	24.08	23.04	22.63
Al_2O_3	5.49	5.47	4.79	4.80
Fe_2O_3	3.90	3.68	3.31	2.77
MgO	1.05	1.08	1.02	1.04
CaO	59.65	58.70	59.84	66.23
SO_3 tot	0.06	4.44	4.43	0.03
LSF	70.4	77.0	82.7	93.5
SR	2.88	2.63	2.84	2.99
AR	1.41	1.49	1.45	1.73

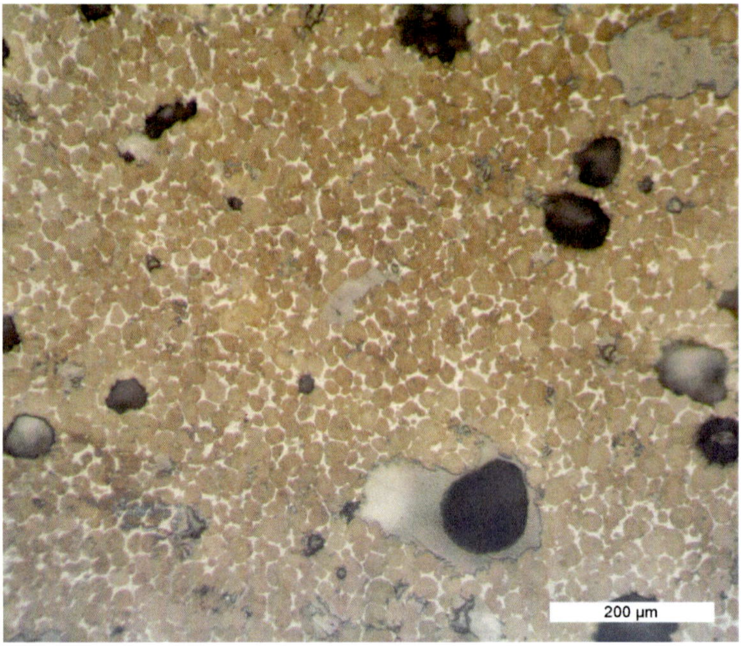

Figure 9. Microstructure of belite clinker S-B80S4 – belite forms brownish grains, the interstitial clinker phases are white, pores are beige (when filled with epoxy) or black (when hollow). Reflected light, dyed by fumes of acetic acid.

The chemistry of belite grains and alite crystals was determined by electron microanalysis. The results are given in Table 9 and 10. The calculations of atomic ratios of elements filling the Ca site (i.e., Ca, Mg, Na, K) versus elements filling the Si site (Si, S, P, Al, Fe^{3+}, Ti) have shown that they are close to stoichiometric 2 : 1 (in belite) and 3 : 1 (in alite).

Figure 10. Microstructure of belite clinker S-B90S4 – alite forms scarce blue crystals, small orange inclusion in alite in the upper right corner – free lime. Reflected light, dyed by acetic acid fumes.

Table 9. The average chemical composition of belite grains determined by electron microprobe [wt. %]

Clinker	S-B	S-B80S4	S-B90S4	S-A
SiO_2	32.008	30.560	28.797	31.012
TiO_2	0.213	0.216	0.219	0.276
Al_2O_3	0.962	1.813	2.759	1.618
Fe_2O_3	1.136	1.170	1.757	1.481
P_2O_5	0.093	0.098	0.093	0.067
MnO	0.050	0.056	0.072	0.073
MgO	0.383	0.246	0.607	0.328
CaO	63.231	63.743	65.362	63.721
Na_2O	0.261	0.107	0.150	0.366
K_2O	0.794	0.102	0.138	0.927
SO_3	0.060	1.614	3.093	0.068
CaO:SiO_2 (atomic)	2.117	2.235	2.432	2.201
(Ca+K+Na+Mg):(Si+S+P+Al+Fe+Ti) (atomic)	2.037	1.970	1.987	2.057

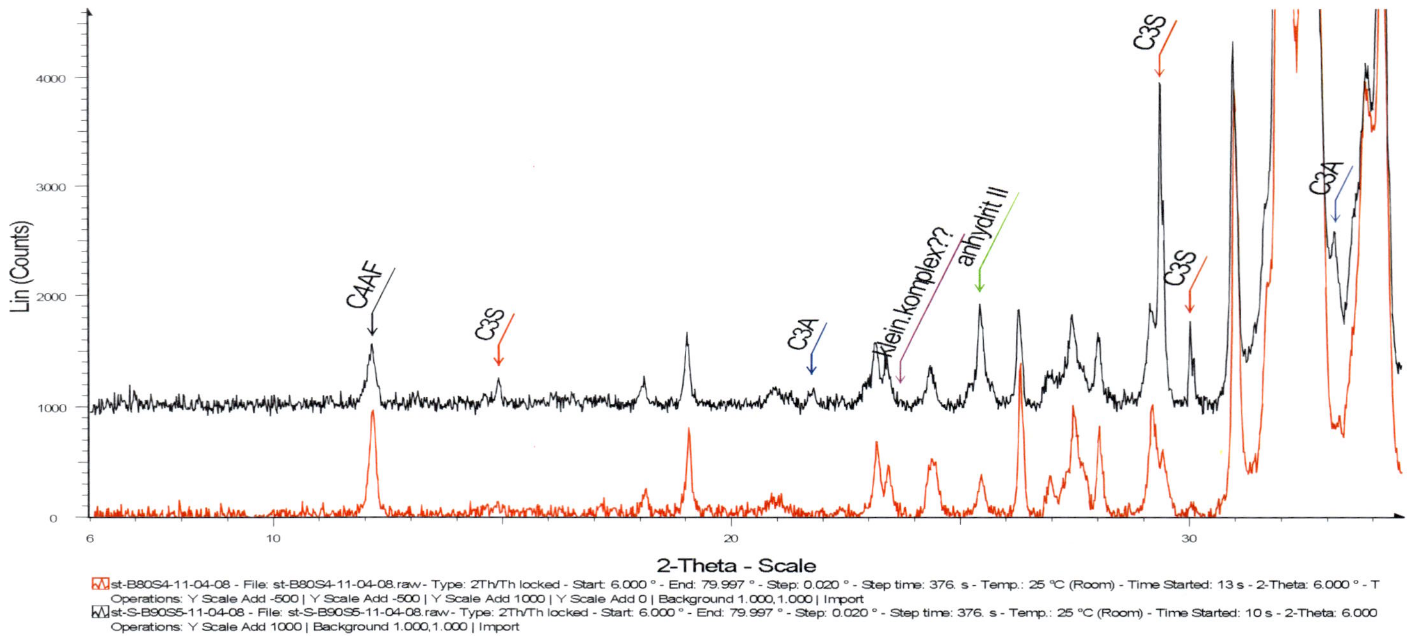

Figure 11. Comparison of XRD patterns of clinkers S-B80S4 (red curve) and S-B90S4 (black curve).

The X-ray diffraction analysis revealed in belite clinkers doped with SO_3 besides the common clinker phases also the presence of anhydrite II ($CaSO_4$). The Klein's complex (mineralogically yeelimite – $Ca_4Al_6(SO_4)O_{12}$) is probably present, but only in negligible amount. The comparison of XRD patterns is presented on Figure 11.

Table 10. The average chemical composition of belite grains determined by electron microprobe [wt. %]

Clinker	S-B90S4	S-A
SiO_2	23.500	25.028
TiO_2	0.140	0.133
Al_2O_3	1.615	0.739
Fe_2O_3	1.065	0.719
P_2O_5	0.084	0.080
MnO	0.067	0.048
MgO	1.127	0.652
CaO	71.810	72.213
Na_2O	0.036	0.096
K_2O	0.016	0.078
SO_3	0.165	0.035
$CaO:SiO_2$ (atomic)	3.056	2.885
(Ca+K+Na+Mg):(Si+S+P+Al+Fe+Ti) (atomic)	2.970	2.952

Cement Properties

The specific surface (Blaine) and specific weight of prepared cements are given in Table 11.

Table 11. Specific weight [kg/m^3] and specific surface [m^2/kg] of prepared cements

Cement	C-B	C-B80S4	C-B90S4	C-A
Specific weight	3230	3261	3191	3179
Specific surface	435	436	438	437

Table 12. Tensile and compressive strengths of cements in MPa

Cement	2 days		7 days		28 days		56 days		90 days	
	Tens.	Com.	Tens.	Com.	Tens.	Com.	Tens.	Com.	Tens.	Com.
C-B	1.8	2.0	1.3	2.3	3.9	14.6	5.1	31.1	5.7	38.3
C-B80S4	0.4	1.8	2.3	12.2	6.9	47.4	8.7	57.8	6.3	60.6
C-B90S4	2.8	11.0	3.4	24.2	7.4	51.1	9.4	61.5	8.7	65.8
C-A	3.7	16.6	7.3	44.7	8.4	66.3	7.9	70.8	8.3	69.5

The principal technological property of cements is their strength; it was measured according to EN 196-1 and the results are given in Table 12. The compressive strengths are presented also graphically on Figure 12.

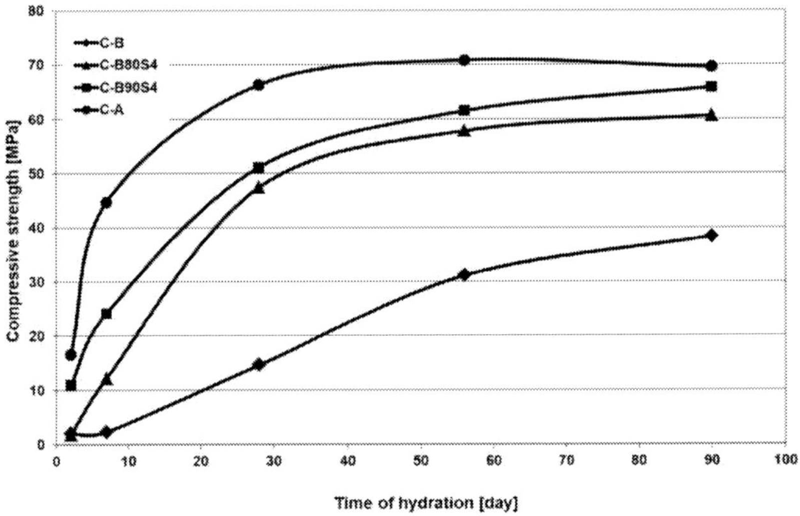

Figure 12. The development of compressive strength in prepared cements.

The results of normal consistency, setting and soundness of cements after EN 196-3 are given in Table 13.

Table 13. Normal consistency, setting and soundness of cements

Cement	Normal consistency [%]	Start of setting [h:min]	Setting time [h:min]	Soundness [mm]
C-B	30.7	1:00	1:30	0.0
C-B80S4	30.3	4:20	5:40	0.8
C-B90S4	28.7	4:00	5:10	0.3
C-A	28.0	3:50	4:40	0.2

Table 14. Hydration heat of cements determined according to EN 196-8 [$J.g^{-1}$]

Cement	2 days	7 days	28 days	56 days	90 days
C-B	118	140	237	286	325
C-B80S4	120	154	286	312	327
C-B90S4	174	195	290	323	346
C-A	230	311	449	475	487

The released hydration heat after 2, 7, 28, 56, and 90 days of hydration, determined by the dissolution method is given in Table 14. Table 15 gives the temperatures of maximum (Tmax) and time needed to reach this temperature (tmax) on the hydration curve from the semiadiabatic calorimeter and the specific hydration heat Q of the cements during 24 hours, calculated from this curve. The course of hydration of cement pastes registered by semiadiabatic calorimetry is presented also graphically on Figure 13.

Table 15. The course of cement pastes hydration obtained by semiadiabatic calorimetry

Cement	Tmax [°C]	Tmax [hours]	Q_{24} [$J.g^{-1}$]
C-B	31.67	2.56	3.2
C-B80S4	26.67	0.19	3.7
C-B90S4	30.24	12.04	71.0
C-A	47.91	10.42	268.9

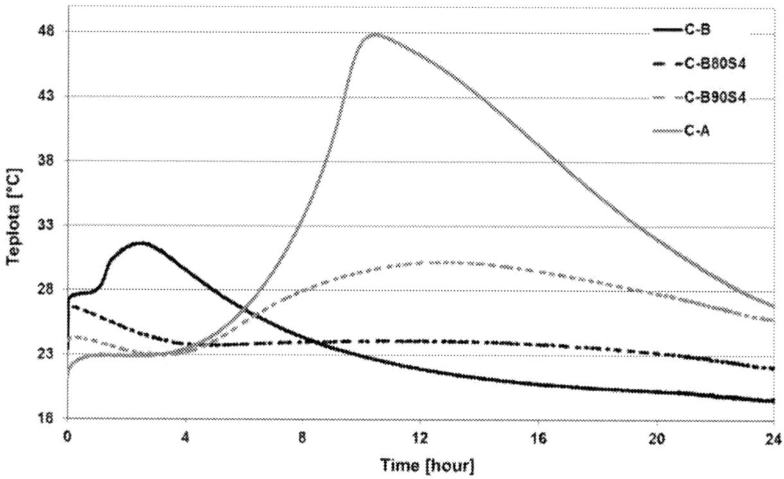

Figure 13. The course of hydration of cement pastes registered by semiadiabatic calorimetr.

Discussion and Results

The entry of SO_3 into the belite structure, where it substitutes SiO_2 facilitates the entry into the belite structure also to further oxides, especially to Al_2O_3 and Fe_2O_3. Due to it, the CaO : SiO_2 ratio in belite structure considerably increases in comparison to belite of clinker without SO_3 addition. This phenomenon is more pronounced in belite clinker with higher lime saturation. Nevertheless, taking into account all elements that were analyzed by electron microprobe and can substitute some of the two main constituents of belite (Ca, Si), the chemistry of belite occurred to be stoichiometric.

Belite clinker doped with SO_4^{2-} anions ("sulfobelite" clinker) is in comparison with common belite clinker markedly more hydraulically activated and the cement prepared of it has technological properties (with the exception of early strengths) close to those of Portland cement with dominant alite. Moreover, this belite cement has distinctly decreased hydration heat, what can be advantageous in certain applications.

Cements prepared from these "sulfobelite" clinkers have distinctly better strength characteristics than cements made of sulfoferroaluminate belite clinker and high-Fe belite clinker, that were obtained in a pilot-plant test in Romania (Popescu, Muntean and Sharp, 2003). Cement from clinker S-B90S4, containing 8.5 wt. % of alite has the same strength parameters as cement made of clinker very high in belite, industrially produced in China (Sui et al., 2005), which contains 20 – 30 wt. % of alite.

Optimization of "Sulfobelite" Cements

The optimization of "sulfobelite" cements was done through a series of experiments, varying the chemical composition of raw meals and the burning regime with the aim of achieving as high as possible strengths in all hydration times.

To prepare the raw meals, common cement-making raw materials were used; the source of SO_3 was industrial FDG product ("energogypsum").

The fundamental characteristics of chosen clinkers and belite cements prepared from them are given in Table 16, together with reference cements.

For comparison, the first three columns give parameters of clinkers prepared without SO_3 addition (B – purely belitic, B81S0 – belitic with 25% alite content and A97S0 – common high alite Portland clinker). The phase composition of clinkers was determined by a combination of optical microscopy and XRD. The cement strengths were determined according to the EN 196-1 standard after 2, 7, 28, and 90 days of hydration; for clarity the data are presented also graphically in Figure 14.

Table 16. Properties of cements prepared from belite clinkers doped with SO_3 and reference cements

Cement	B	B81S0	A97S0	B80S4	B92S8 II	B92S6	B89S5 II
Phase	Phase composition of used clinker [wt. %]						
C_3S	0.2	24.5	75.8	0.0	10.0	33.6	16.6
C_2S	79.3	58.4	6.9	81.3	71.5	50.8	70.3
C_3A	4.0	6.5	11.0	2.6	3.7	4.0	2.1
C_4AF	16.5	10.0	5.8	14.6	11.4	8.8	8.8
C_{free}	0.0	0.5	0.5	0.0	1.9	0.8	1.4
$C_4A_3\overline{S}$	0.0	0.0	0.0	traces	0.0	0.0	0.0
$C\overline{S}$	0.0	0.0	0.0	1.5	1.5	2.0	0.8
Parameter	Chemical parameters of used clinker [wt. %]						
SO_3 (wt.%)	0.06	0.07	0.02	4.44	6.77	5.48	4.84
LSF (%)	70.4	79.9	96.2	77.0	89.4	92.1	86.0
SR	2.88	3.00	2.57	2.63	3.12	3.31	3.22
AR	1.41	1.28	1.41	1.49	1.45	1.33	1.58
Parameter	Clinker burning parameters [°C, min]						
Temperature	1400	1350	1450	1400	1350	1350	1350
Burning time	40	50	120	40	30	60	60
Parameter	Cement parameters [wt. %, kg/m^3, m^2/kg]						
Gypsum addition	4.0	3.0	4.0	4.0	3.0	3.0	2.0
Specific weight	3230	3225	3172	3261	3231	3191	3239
Specific surface	435	397	398	436	402	410	399
Hydration time	Compressive strength [MPa]						
2 days	2.0	4.5	21.6	1.8	16.9	31.7	24.3
7 days	2.3	12.8	49.3	12.2	26.1	51.6	47.1
28 days	14.6	63.5	66.9	47.4	36.4	64.7	64.4
90 days	38.3	71.2	68.8	60.6	64.9	76.2	70.4

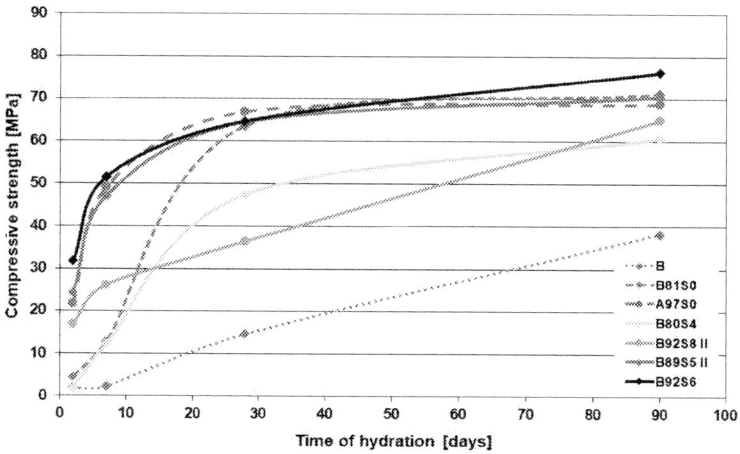

Figure 14. The changes in compressive strength of studied cements.

Table 17 gives the values of released hydration heat for individual cements after 7 days of hydration, determined by the dissolution method according to EN 196-8.

Table 17. The heat of hydration of cements determined according to EN 196-8 [kJ/kg]

Cement	Heat of hydration
B	140
B81S0	232
A97S0	413
B80S4	154
B92S8 II	211
B92S6	305
B89S5 II	257

Along with it, the course of temperature changes during hydration of cement pastes (Figure 15) was observed in semiadiabatic calorimeter (Brandštetr et al., 2001), meeting the EN 196-9 standard.

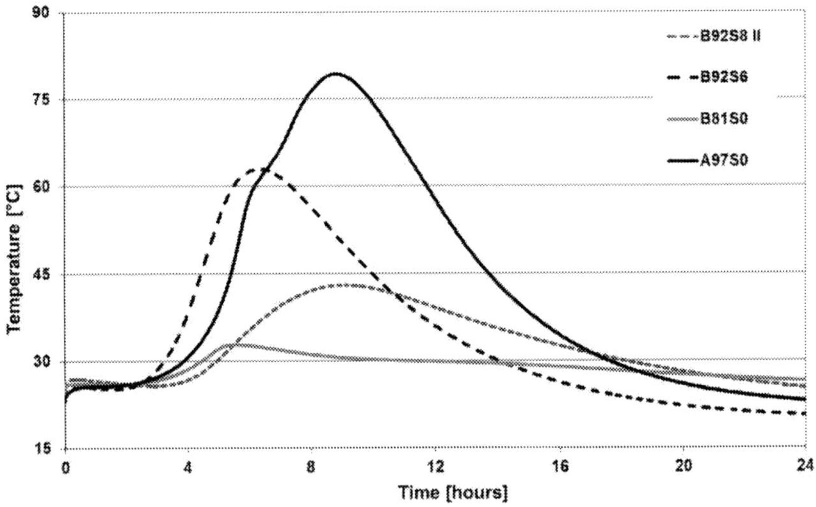

Figure 15. The course of hydration of cement pastes, recorded by semiadiabatic calorimetry.

The results show that belite clinker doped with SO_3 (so called sulfobelite clinker) is in contrast to common belite clinker distinctly more hydraulically active and that cement prepared of it verges with its technological properties to the Portland cement with prevailing alite. At alite content around 20 wt. % and burning temperature of 1350°C it is possible to achieve at common fineness of grinding compressive strength of 20 MPa after 2 days hydration.

Very favorable for certain applications is also the decreased hydration heat of these cements, which can therefore be classified as low hydration heat cements with hydration heat below 270 kJ/kg and some even in the category of cements with very low hydration heat (below 220 kJ/kg).

Optimization of the Parameters According to EN 197-1

The highly belite sulphate-activated cement with similar technological parameters to conventional Portland cement has developed in laboratory (Staněk and Sulovský, 2015). In the last phase, however, its composition had to be optimized to fully meet the requirements of EN 197-1, especially in content of SO_3 content and short-term strength.

Solving

The raw materials used for the preparation of experimental raw meals were common cement-making materials – pure limestone (L1), limestone with increased SiO_2 content (L2), clay shale (CS), iron correction (Fe) and FGD gypsum (EG). The composition given in Table 18 was chosen based on the chemical composition of the raw materials and the performance of experimental clinker burnings with different chemical parameters.

Table 18. The composition of chosen raw meal in wt. % and its basic chemical parameters

Component	B90.5S4
L1	52.05
CS	20.43
L2	21.11
Fe	0.81
EG	5.60
Total	100.00
LSF	90.5
SR	3.0
AR	1.6

Based on the results of the control burning of this raw meal (see Table 19), 3 kg of raw meal was composed according to Table 18 and ground to a rest of 15 wt. % on 0.09 mm sieves.

The raw meal was pressed into tablets 4 cm in diameter, weighing about 100 g and burned at 1350°C for 75 minutes.

The phase composition of the prepared clinker is shown in Table 20, the chemical analysis in Table 21, the basic chemical parameters in Table 22 and clinker microstructure is visible in Figure 16.

Table 19. Phase composition of control clinker B90.5S4

Phase	XRD [%]	Microscopy [wt. %]
C_3S	27.6	25.2
C_2S	59.9	60.7
C_3A cubic	4.0	5.2
C_3A orthorombic	0.2	
C_4AF	7.7	7.0
C_{free}	0.6	1.8

Table 20. Phase composition of clinker B90.5S4 for cement preparation

Phase	XRD [%]	Microscopy [wt. %]
C_3S	23.0	19.7
C_2S	62.4	65.2
C_3A cubic	5.0	5.8
C_3A orthorombic	0.3	
C_4AF	8.3	7.6
C_{free}	0.6	1.8

Table 21. Chemical analysis of clinker B90.5S4 in wt. %

Component	B90.5S4
LOI	0.85
SiO_2	23.10
CaO	61.90
TiO_2	0.26
P_2O_5	0.06
Na_2O	0.29
K_2O	0.96
MgO	1.06
MnO	0.09
Fe_2O_3	3.13
Al_2O_3	5.25
SO_3	2.83

Table 22. Basic chemical parameters of clinker B90.5S4

Parameter	Value
LSF	84.9
SR	2.76
AR	1.68

Figure 16. Microstructure of clinker B90.5S4 – blue crystals of alite, brown grains of belite and violet grains of free lime. Reflected light, dyed by acetic acid fumes.

The clinker with the addition of pure gypsum was ground in a laboratory ball mill and its basic characteristics are given in Table 23.

Determination of normal consistency, setting and soundness was performed according to EN 196-3 and the results are shown in Table 24.

Table 23. Parameters of prepared cement

Cement	Addition of gypsum [wt. %]	Content of SO_3 [wt. %]	Specific surface [m^2/kg]
CEM B90.5S4	2	4.0	402

Table 24. Normal consistency, setting and soundness of prepared cement

Cement	Normal consistency [%]	Start of setting [min]	Setting time [min]	Soundness [mm]
CEM B90.5S4	29.7	150	210	1.3

Determination of compressive and tensile strength of cement was performed on standard cement mortars according to EN 196-1after 2, 28 and 90 days of hydration (see Table 25). A comparison of cement parameters with the requirements of EN 197-1 is given in Table 26.

Table 25. Strengths of cement B90.5S5 according to ČSN EN 196-1 in MPa

Cement	2 dny		28 dní		90 dní	
	Tens.	Com.	Tens.	Com.	Tens.	Com.
CEM B90.5S4	4.4	20.5	8.1	50.0	7.8	57.8

Table 26. Parameters of prepared cement

Parameter	EN 197-1 Class 42.5 R	CEM B90.5S4	Evaluation
LOI [wt. %]	≤ 5.0	1.91	complying
Insoluble residue [wt. %]	≤ 5.0	0.14	complying
Content of sulfate (SO_3) [wt. %]	≤ 4.0	4.0	complying
Content of chloride [wt. %]	≤ 0.01	0.007	complying
Compressive strength 2 days [MPa]	≥ 20.0	20.5	complying
Compressive strength 28 days [MPa]	≥ 42.5 ≤ 62.5	50.0	complying
Start of setting [min]	≥ 60	150	complying
Soundness [mm]	≤ 10	1.3	complying

Conclusion

The research and development of special low-energy cements is very topical issue and their practical mass use would have considerable environmental and economic effects, particularly a decrease of CO_2 emissions, lower consumption of energy for burning and lower extraction of natural materials.

The long-standing experience in use of common Portland cement and its time-tested guaranteed durability, however, obstruct their mass utilization and they are used mainly for special purposes.

The most important producer of these cements is China; over 1 million tons of these cements are made for both structural and non-structural purposes every year (Janotka, Krajči and Mojumdar, 2007).

The presented cement prepared in the laboratories of the Research Institute for Building Materials represents a new type of hydraulically active low-energy belite cement.

The clinker for its preparation, containing only 20 to 30 wt. % of alite, was burned at a temperature of 1,350°C and is activated by the addition of SO_3. Compared to SAB cements it does not contain $C_4A_3\overline{S}$. It contains a small proportion of anhydrite.

Chemical activation of belite by sulfate anions has significant impact on its hydraulic activity. The incorporation of SO_3 into the belite structure, where it substitutes SiO_2, enables to widen this substitution also for other oxides, above all Al_2O_3 a Fe_2O_3 (Staněk and Sulovský, 2015). In this way, the CaO : SiO_2 ratio in belite is distinctly higher than in belite without the SO_3 incorporation. This phenomenon is more pronounced in belite clinker with higher lime saturation factor.

Cements prepared from these clinkers have at the same fineness of grinding technological parameters equal to those of alite-rich cement – including short-term strengths, which exceed 20 MPa. In addition, they have a favourably lower heat of hydration after 7 days of hardening.

In contrast to purely belite clinker doped with SO_3, a small amount of alite distinctly supports the growth of early strengths, because this alite is present in its more hydraulically active M_1 modification (Staněk and Sulovský, 2002), which is stabilized just by SO_3.

The relatively high lime saturation (yet still about 10% lower than in common Portland clinker) in this belite clinker ensures at the reaction with water formation of higher amounts of portlandite ($Ca(OH)_2$) and increases the overall alkalinity, accelerating thus the course of hydration.

Hydraulic activation of this "sulfobelite" clinker is therefore caused on one hand by the disturbance of the belite crystal structure and its "contamination" by foreign ions, on the other hand by relatively high CaO content in clinker.

This activated belite clinker could be used with regard to its good hydraulic properties for the production of self-contained "sulfobelite" cement, or for blending with OPC for production of special cements. In working practice, it would be advantageous to use the existing kiln lines for alternating production of "sulfobelite" and classical alite clinker. The lower burning temperature and very short clinkering zone during the belite clinker burning would enable to use the sticking from the preceding burning of the alite clinker, because the sticking from belite burning is probably insufficiently stable.

The increase in lime saturation causing formation of up to 30 wt. % alite would contribute to the increase of hydraulic activity as well as to improvement of parameters of burning and grindability. Granules with prevalent belite are plastic and have higher coverage of their surface by the melt (closer packing of spheroidal belite grains that do not coalesce), while the coalescing angular crystals of alite form a solid skeleton of the granule with lower coverage of the surface by the melt and the burning becomes safer for the kiln.

Introduction of this cement into the cement making industry would lead to a decrease in energy consumption for production, a saving of high-quality limestone, utilization of waste raw materials with high SO_3 content and at the same time the lowering of CO_2 emissions.

In practice, it is very important that in the final phase of the solution, this low-energy cement was prepared in such a way that its parameters comply with CEM I cement of strength class 42.5 R according to EN 197-1.

Acknowledgements

This research was done with the Institutional support for long-term development of research organizations by the Ministry of Industry and Trade of the Czech Republic.

References

Abdul-Maula, S., Odler, I. (1992). SO_3-rich Portland cements: synthesis and strength development. *Materials Research Society Symposium Proceedings* 245, 315-320.

Ali, M. M., Sharma, P. S., Bhargava, R. (1992). Hydration characteristics of magnesia assimilated sulfoaluminate-belite cement. *9th International Congress on Chemistry of Cement*, New Delhi, September 1992, Vol. 3, 227-232.

Andac, M., Glasser, F.P. (1995). Microstructure and microchemistry of calcium sulfoaluminate cement. *Materials Research Society Symposium Proceedings* 370, 135-142.

Beretka, J., Sherman, N., Marroccoli, M., Pompo, A., Valenti, G. L. (1997). Effect of composition on the hydration properties of rapid hardening sulfoaluminate cement. *10th International Congress on Chemistry of Cement*, Gothenburg, 2-6 June 1997, paper 2ii029.

Bigare, M., Guinier, A., Mazieres, C., Regourd, M., Yannaquis, N., Eysel, W., Hahn, T., Woermann, E. (1967). Polymorphism of tricalcium silicate and its solid solutions. *Journal of the American Ceramic Society* 50, 609-619.

Bikbaou, M. Y. (1986). Formation, crystal chemistry and properties of alinite and jasmundite. *8th International Congress on Chemistry of Cement*, Rio de Janeiro, 22-27 September 1986, vol. 2, 352-357.

Boikova, A. I., Fomicheva, O. I., Grishchenko, L. V. (1992). Effect of SO_3 and MgO on phase relations and crystal chemical state of phases in belite-aluminiferrite clinkers, *9th International Congress on Chemistry of Cement*, New Delhi, September 1992, Vol. 2, 413-419.

Boikova, A. I., Grishchenko, L. V., Domansky, A. I. (1986). Hydration activity of chlorine containing phases. *8th International Congress on Chemistry of Cement*, Rio de Janeiro, 22-27 September 1986, vol. 3, 275-276.

Brandštetr, J. (1999). Some prospective inorganic composite materials for the 21st century (in Czech), *4th Conference VÚSH*, Brno, 11-12 November, 104-111.

Brandštetr, J., Polcer, J., Krátký, J., Holešinský, R., Havlica, J. (2001). Possibilities of the use of isoperibolic calorimetry for assessing the hydration behavior of cementitious systems. *Cement and Concrete Research* 31, 941-947.

Brown, A. D. R. (1992). Commercial, production, composition and properties of a calcium sulfoaluminate cement. *International Conference on Cements for the Future - Calcium sulphoaluminates*, London, 3.

Bullerjahn, F., Schmitt, D., Ben Haha, M., Batog, B., Zajac, M. (2012). Calcium sulfoaluminate cement with ternesite. *US Patent* 20140230697.

Chatterjee, A. (1996). High Belite Cements–present status and future technological option. (Part I) *Cement and Concrete Research* 26, 1213-1225.

Christensen, N. H., Jepsen, O. L., Johansen, V. (1978). Rate of alite-formation in clinker sandwiches. *Cement and Concrete Research* 8, 693-701.

Chromý, S. (1974a). Mechanism of white clinker formation. *6th International Congress on Chemistry of Cement*, Moscow, September 1974, Vol. III, 268-271.

Chromý, S. (1974b). Anfärben des freiem CaO und Silikate in anschliffen von Portlandklinker [*Staining of the free CaO and silicates in sanded Portland clinker*]. *Zement-Kalk-Gips* 27, 79-84.

Chromý, S. (1978). Accuracy and precision of microscopic quantitative phase analysis of Portland clinkers. (in Czech). *Silikáty* 22, 215-226.

Chromý, S. (1982) Process der Portlandzemententstehung, Reaktivität und Brennbarkeit von Zementrohstoffen Teil I. [*Process of Portland cement formation, reactivity and flammability of cement raw materials. Part I*]. *Zement-Kalk-Gips* 35, 204-210.

Chromý, S., Hrabě, Z. (1982). Process der Portlandzemententstehung, Reaktivität und Brennbarkeit von Zementrohstoffen. Teil II. [*Process of Portland cement formation, reactivity and flammability of cement raw materials. Part II*]. *Zement-Kalk-Gips* 35, 368-373.

Chrysafi, R., Perraki, T., Kakali, G. (2007). Sol-gel preparation of $2CaO.SiO_2$, *Journal of the European Ceramic Society* 27, 1707-1710.

Chvatal, T. (1972). New rapid setting cement. (in German) *Zement-Kalk-Gips* 26, 385-391.

Dreizler, I., Knöfel, D. (1982). Der Eifluss von Magnesiumoxid auf die Zementeigenschaften [*The egg flow of magnesium oxide to cement properties*]. *Zement-Kalk-Gips* 35, 537-550.

Emanuelson, A., Landa-Cánovas, A. R., Hansen, S. (2003). A comparative study of ordinary and mineralised Portland cement clinker from two different production units. Part II: Characteristics of the calcium silicates. *Cement and Concrete Research* 33, 1623-1630.

Feng, X., Zhu, Y. (1986). Research on an early strength cement containing high content of iron. *8th International Congress on Chemistry of Cement*, Rio de Janeiro, 22-27 September 1986, Vol. 2, 286-292.

Fukuda, K., Maki, I., Ito, S. (1992). Remelting reaction within belite crystals during cooling. *Journal of the American Ceramic Society* 75, 2896-2898.

Fukuda, K., Wakamatsu, N., Ito, S. (2001). Improvement in hydration reactivity of alpha-phase belite by remelting reaction. *Journal of the American Ceramic Society* 84, 639-641.

Ghosh, S. H., Bhaskara Rao, P., Paul, A.K., Raina, K. (1979). The chemistry of dicalcium silicate mineral. *Journal of Materials Science* 14, 1554-1566.

Gies, A., Knöfel, D. (1986). Influence of alkalis on the composition of belite-rich cement clinkers and the technological properties of the resulting cements. *Cement and Concrete Research* 16, 411-422.

Gies, A., Knöfel, D. (1987). Influence of sulfur on the composition of belite-rich cement clinkers and the technological properties of the resulting cements. *Cement and Concrete Research* 17, 317-328.

Grosvalds, I., Lagzdina, S., Sedmalis, U. (1997). Dolomitic roman cement: a low temperature hydraulic binder. *10th International Congress on Chemistry of Cement*, Gothenburg, 2-6 June 1997, paper 2ii009.

Hewlett, P. C. (2007). *Lea's Chemistry of Cement and Concrete.* 4th Edition, Elsevier.

Hlaváč, J. (1988). *Foundations Technology of Silicates.* (in Czech), SNTL/Alfa, Praha.

Huang, W., Qian, G., Yang, Y., Qi, Y. (1992a). New energy-saving cement – iron-rich fluoro-aluminate cement. *9th International Congress on Chemistry of Cement*, New Delhi, September 1992, Vol. 3, 284-290.

Huang, W., Yang, Y., Qian, G., Qi, Y. (1992b). The research on $C_2A_xF_{(1-x)}$ iron-rich fluoro-aluminate cements. *9th International Congress on Chemistry of Cement*, New Delhi, September 1992, Vol. 3, 9-15.

Ichikawa, M., Kanaya, M. Effects of minor components and heating rates on the fine textures of alite in Portland cement clinker. (1997). *Cement and Concrete Research* 27, 1123-1129.

Ikebata, T., Miyama, Y., Sato, M. (1994). The effects of clinkering temperature and silicate phase on the hydration of $C_{11}A_7.CaF_2$. *Cement Technology AcerS*, 27-34.

Ilyukhin, V. V., Nevsky, N. N., Bickbau, M. J., Howie, R. A. (1977). Crystal structure of alinite. *Nature* 269, 397-398.

Janotka, I., Krajči, L., Mojumdar, S. C. (2007). Performance of sulphoaluminate-belite cement with high $C_4A_3\overline{S}$ content. (online) homepage: https://www.ceramics-silikaty.cz/2007/pdf/2007_02_74.pdf (date of access: 2022-01-23).

Ji, L., Ren, X., Su, H. (1997). Effect of chloride on hydration properties of alinite cement. *10th International Congress on Chemistry of Cement*, Gothenburg, 2-6 June 1997, paper 2ii030.

Johansen, V. (1979). Anwendung des Gleichgewichtdiagrams beim Klinkerbindung-vorgang industriell herstelter Klinker [*Application of the equilibrium diagram in the clinker binding process of industrially manufactured clinker*]. *Zement-Kalk-Gips* 32, 176-181.

Jost, K. H., Seydel, R., Müller, A., Stark, J. (1988). Relationship between phase-composition, cooling rate and Na_2O content in belite clinkers. *Zement - Kalk - Gips* 41, 169-170.

Jost, K. H., Ziemer, B. (1984). Relations between the crystal structures of calcium silicates and their reactivity against water. *Cement and Concrete Research* 14, 177-184.

Knöfel, D., Wang, J. F. (1994). Properties of three newly developed quick cements. *Cement and Concrete Research* 24, 801-812.

Kondo, R., Choi, S. (1968). Mechanism and kinetics of Portland cement formation for an example of the solid state reaction in the presence of a liquid phase. *5th International Symposium on the Chemistry of Cement*, Tokyo, 7-11 October 1968, Part I, 163-171.

Kozłowski, R., Hughes, D., Weber, J. (2010). Roman cements: key materials of the built heritage of the 19th century. In: *Materials, Technologies and Practice in Historic Heritage Structures*, Springer Netherlands, 259-277.

Kurdowski, W., Duszak, S., Trybalska, B. (1997). Belite produced by means of low-temperature synthesis. *Cement and Concrete Research* 27, 51-62.

Lampe, F., Hilmer, W., Jost, K. H., Reck, G., Boikova, A. I. (1986). Synthesis, structure and thermal decomposition of alinite. *Cement and Concrete Research* 16, 505-510.

Liu, K., Deng, Z., Yuan, M, Tang, J. (1992). Study on cement in the $C_{11}A_7.CaF_2$ system. *9th International Congress on Chemistry of Cement*, New Delhi, September 1992, Vol. 3, 268-276.

Ludwig, U., Urbonas, L. (1993). Synthesis and reactivity of fluoralinites and fluoralinite cement. (in German). *Zement - Kalk - Gips* 46, 568 – 572.

Maki, I., Chromý, S. (1978a). Microscopic study on the polymorphism of Ca_3SiO_5. *Cement and Concrete Research* 8, 407-414.

Maki, I., Chromý, S. (1978b). Characterization on the alite phase in Portland cement clinker by microscopy. *Il Cemento* 75, 247-252.

Maki, I., Fukuda, K., Seki, S., Taioka, T. (1991). Impurity distribution during crystal growth of alite in Portland cement clinker. *Journal of the American Ceramic Society* 74, 2082-2085.

Maki, I., Fukuda, K., Yoshida, H., Kumaki, J. (1992). Effect of MgO and SO3 on the impurity concentration in alite in Portland cement clinker. *Journal of the American Ceramic Society* 75, 3163-3165.

Maki, I., Goto, K. (1982). Factors influencing the phase constitution of alite in Portland cement clinker. *Cement and Concrete Research* 12, 301-308.

Mehta, P. K. (1980). Investigations on energy saving cements. *World Cement Technology* 11, 166-177.

Midgley, H. G., Bhaskara Rao, P. (1978). Formation of stratlingite, $2CaO.SiO_2.Al_2O_3.8H_2O$, in relation to the hydravion of high alumina cement. *Cement and Concrete Research* 8, 169-172.

Mojumdar, S. C., Janotka, I. (2002). Thermophysical properties of blends from Portland and sulfoaluminate-belite cements. *Acta Physica Slovaca* 52, 435-446.

Morsli, K., De la Torre, A. G., Zahir, M., Aranda, M. A. G. (2007). Mineralogical phase analysis of alkali and sulfate bearing belite rich laboratory clinkers. *Cement and Concrete Research* 37, 639-646.

Noudelman, B. I., Gadaev, A. I. (1986). Physico-chemical aspects of the crastalization of chlorsilicates after clinkerization at low temperatures in melted salts. (in French), *8th International Congress on Chemistry of Cement*, Rio de Janeiro, 22-27 September 1986, vol. 2, 347-351.

Odler, I. (2000). *Special Inorganic Cements.* E & FN Spon, London.

Odler, I., Abdul-Maula, S. (1980). Structure and properties of Portland cement clinker doped with CaF_2. *Journal of the American Ceramic Society* 63, 654-659.

Osokin, A. P., Krivoborodov, J. R., Dyukova, N. F. (1992). Sulfoferrite cements. *9th International Congress on Chemistry of Cement*, New Delhi, September 1992, Vol. 3, 256-261.

Osokin, A. P., Potapova, E. N. (1989) Alit-Bildung in Oxid-Salz-Schmelzen [*Alite formation in oxide-salt melts*]. *Silikattechnik* 40, 4-6, 149-153, 229-234.

Popescu, C. D., Muntean, M., Sharp, J. H. (2003). Industrial trial production of low energy belite cement. *Cement and Concrete Composites* 25, 689-693.

Pradip, A., Kapur, P. C. (1990). Production and properties of alinite cements from steel plant waste. Cement and Concrete Research 20, 15-24.

Puertas, F., Varela, M. T. B., Molina, S. G. (1995) Kinetics and thermal decomposition of $C_4A_3\overline{S}$ in air. *Cement and Concrete Research* 25, 572-580.

Quillin, K. (2001). Performance of belite-sulfoaluminate cements. *Cement and Concrete Research* 31, 1341-1349.

Sahu, S. (1994). Developments in low energy clinkers. *Ceramics-Silikáty* 38, 191-200.

Santoro, L., Garofano, R., Valenti, G. K. (1986). Calcium sulphoaluminate made from phosphogypsum and its hydration properties. *8th International Congress on Chemistry of Cement*, Rio de Janeiro, 22-27 September 1986, Vol. 4, 389-394.

Sharara, A. M., El-Didamony, H., Ebied, E., El-Aleem, A. (1994). Hydration characteristic of β-C_2S in the presence of some pozzolanic materials. *Cement and Concrete Research* 24, 966-974.

Staněk, T. (2014). Influence of preparation parameters on properties of belite clinker. *Advanced Materials Research* 897, 3-8.

Staněk, T. (2016). The influence of SO_3 and MgO on kinetics of alite formation. *Procedia Engineering* 151, 26 – 33.

Staněk, T., Sulovský, P. (2002). The influence of the alite polymorphism on the strength of the Portland cement. *Cement and Concrete Research* 32, 1169-1175.

Staněk, T., Sulovský, P. (2015). Active low-energy belite cement. *Cement and Concrete Research* 68, 203-210.

Strunge, J., Knöfel, D., Dreizler, I. (1985, 1986). Einfluss der Alkalien und Schwefels auf die Zementeigenschaften. Teil I [*Influence of alkalis and sulfur on cement properties. Part I*] *Zement-Kalk-Gips* 38, 150-158. Teil II [*Part II*]. *Zement-Kalk-Gips* 38, 441-450. Teil III [*Part III*]. *Zement-Kalk-Gips*. 39, 380-456.

Sui, T., Fan, L., Wen, Z., Wang, J., Zhang, Z. (2004). Study on the properties of high strength concrete using high belite cement. *Journal of Advanced Concrete Technology* 2, 201-206.

Sui, T., Liu, K., Guo, S. (1998). Research on high belite cement. Part I and Part II. *4th Beijing International Symposium on Cement and Concrete*, Beijing, China, 27-30 October 1998, Vol. 2, 27-30.

Sui, T., Wen, Z., Wang, J., Fan, L. (2005). Development of belite based cements in China. Cement Combinations for Durable Concrete. *International Conference held at the University of Dundee*, Dundee, 5-7 July 2005, 323-328.

Taylor, H. F. W. (1999). Distribution of sulfate between phases in Portland cement clinkers. *Cement and Concrete Research* 29, 1173-1179.

Uchikawa, H. (1994). Management and strategy in cement technology for the next century. *World Cement Technology* 25, 66-78.

Uchikawa, H., Tsukiyma, K. (1973). The hydration of jet cement at 20°C. *Cement and Concrete Research* 3, 263-277.

Viswanathan, V. N., Raina, S. J., Chatterjee, A. K. (1978). An exploratory investigation on porsal cement. *World Cement Technology* 9, 109-118.

Wang, J. F., Knöfel, D. (1994). Properties of three newly developed quick cements. *Cement and Concrete Research* 24, 801-812.

Wang, Y., Deng, J. (1986). An investigation into cement CaO-SiO_2-Al_2O_3-Fe_2O_3-SO_3 system. *8th International Congress on Chemistry of Cement*, Rio de Janeiro, 22-27 September 1986, Vol. 2, 300-305.

Wang, Y., Su, M. (1989). Recent development of new kinds of cement in China. *2nd Beijing Conference report*.

Weber, M. (1989). Stabilization of Portland clinker burning in a rotary kiln. (in Czech), Brno, *Candidate of Science Thesis*.

Zakharov, L. A. (1974). Alumino-belite cement, *6th International Congress on Chemistry of Cement*, Moscow, September 1974, supple. paper.

Chapter 3

The Use of Synthesized Calcium Silicates in Compositions for the Restoration of Buildings

V. Loganina and K. Sergeeva
Department of Quality Management
and Technology of Building Production,
Penza State University of Architecture and Construction,
Penza, Russia

Abstract

Lime mixes are the primary material for restoration work in historic buildings. The research object of this study is modifying lime binders with specially synthesized calcium silicate hydrates (CSHs). This study aimed to improve lime render mixes' weather resistance. Hydro silicates of calcium (HSC) are received by synthesis from liquid sodium glass in the presence of calcium chloride CaCl2 with the subsequent drying of sediment at a temperature 105-300°C and its crushing. Studying of material composition of the filler, made by methods of the X-ray phase analysis (RFA) and infrared (IK) spectroscopy, and also structure on the electronic scanning microscope of PhenomTM G2 pro showed the existence in synthesizable material the calcium silicates of various basicity.

The conducted research show that the filler based on HSC possesses hydraulic activity. The activity of filler, defined according to a technique [5], makes 195-250 mg/g depending on the synthesis mode.

The following factors were considered: the density of the liquid glass, the amount of the precipitating additive, the rate of introduction of the precipitating additive, the drying mode of the precipitate, and the storage time of the precipitate. The research methods were X-ray

In: Properties and Uses of Calcium Silicate
Editor: Amanda G. Carlton
ISBN: 979-8-88697-128-6

diffraction analysis, differential thermal analysis, Fourier transform infrared spectrometry, and optical and electron microscopy. Regularities of activity change of filler interaction with lime depending on conditions of its receiving are revealed. It was revealed that lime compositions with CSH have a higher strength gain rate than the control compositions. Decrease of amount of free lime in limy samples with additive of synthesizable hydro silicates, increase of durability are revealed. A mathematical model of the kinetics of hardening a lime composite based on a binder filled with CSH was obtained. The regularities of the change in the lime composite's strength depending on the filler grinding's fineness, its content, and the amount of mixing water have been established. It was revealed that the introduction of CSH into the lime composition increases the weather resistance of facade lime mixtures by reducing the porosity and increasing the volume of closed pores of the composite.

Keywords: lime, calcium silicate hydrates, modification, restoration mix

Introduction

For the restoration of buildings and structures of historical development and the decoration of newly erected objects, dry plaster and render mixes are widely used [Dylewski, R.; Adamczyk, J., 2014; De Azevedo, A.R.G., 2017]. The production of dry mixes is one of the most rapidly developing construction industry areas and one of the largest construction market segments [Groot, C, 2009; Murmu, M.2015]. Since 2015, the construction sector has seen an increase in production of dry plaster and render mixes by 10–25% [Turanli, L.; Saritas, A.2011]. The volume of production of plaster, render and putty mixtures is increasing every year, and today it is about 48% of the production of dry mixes [Yardim, Y.2016; Zimich, 2016].

The traditional materials used for the finishing and renovation of facades for many years have been lime compounds [Cerný, 2006; Loganina, 2021]. Lime dry render mixes are used mainly to restore architectural monuments and buildings since restoration of historical buildings causes certain difficulties associated with lime render incompatibility with finishing materials [Cultrone, 2007; Loganina, 2021]. Additionally, lime compositions are characterized by higher fracture toughness compared to cement ones [D'Armada, 2012; Ibragimov, 2019]. In particular, the ratio of flexural strength to compressive strength for cement materials is 0.1–0.15, and for lime compositions, this ratio is 4–6 times higher, amounting to 0.6 [Pavlík, 2016; Loganina, 2017].

However, coatings based on lime-based dry mixes are characterized by low water resistance, which leads to increased costs [Hauptmann, 2000]. In this regard, it is urgent to develop a technological solution that increases coatings' durability based on lime dry mixes [Loganina, 2014; Fediuk, 2019; Popov, 2016].

To increase the water resistance of lime coatings, Pahlavan et al. [Pahlavan, 2017] propose the introduction of hydrophobic additives to the mix formulation, which, having an air entrainment of 3–5%, form a network of closed pores. Accordingly, the stresses arising from the freezing of water in open pores decrease, thereby significantly increasing the coatings' frost resistance [Hamard, 2013; Asaad, 2018; Loganina, 2018].

Another way to improve performance is the use of active mineral additives. In particular, the use of mineral additives leads to an increase in water resistance, impermeability and chemical resistance, an improvement in crack resistance, and an increase in compressive strength [Vejmelková, 2012; Loganina, 2014]. High activity fillers increase the adhesion of solutions based on dry building mixes with the base [Walker, 2010; Zegardło, 2016]. Natural additives (silica, alumina, casting-box, diatomite, ashes, tuffs, pumice, volcanic ash) and artificial additives (granulated blast-furnace slags, fuel ash, and slag, artificially fired clay materials, silica waste) are used [Pogorelov, 2016]. The interaction of lime with active mineral additives is based on the fact that active (amorphous and finely dispersed) silica binds lime in the presence of water [Pu, 1999]. As a result of this reaction, calcium silicate hydrates are formed, determining the hydraulic hardening of lime—i.e., strength build-up under water after preliminary hardening in air [Ezziane, 2009]. In addition to silica, the active mineral additives contain alumina, which in the presence of moisture can also interact with lime, forming calcium hydro aluminate with hydraulic properties [. Inozemtcev, 2018].

The use of calcium silicate hydrate (CSH) additives to enhance the performance of limestone formulations is much less studied.

In this regard, as a scientific hypothesis, an assumption was made about the possibility of increasing the operational resistance of lime finishing coatings by introducing active fillers based on calcium silicate hydrates into dry render mix formulations.

The study aimed to improve lime render mixes' weather resistance by modifying lime binders with specially synthesized calcium silicate hydrates. The following tasks were solved to achieve this goal (next page):

- Studying of the regularities of the synthesis of finely dispersed fillers based on calcium silicate hydrates;
- Researching of the structure formation of lime systems in the presence of CSH fillers.

Method

The initial binder was lime (Atmis-sugar, Russian Federation) with an activity of 75–84%, a real density of 2205 kg/m^3, a bulk density of 500 kg/m^3, and a specific surface area of 10,000 cm^2/g. Liquid sodium glass with a different silicate modulus and calcium chloride $CaCl_2$, which acted as a precipitant, was used to obtain filler based on CSH. The amount of added precipitant was 30–50% by weight of water glass. The precipitant was added as a 7.5–15% water solution, depending on the density and silicate modulus of water glass. A silicate modulus of water glass was calculated as the ratio of the number of grams of silicon dioxide molecules to the number of grams of sodium oxide molecules. To obtain colored compositions, an additive-chromophore $FeCl_3$ was introduced, as well as pigments (blue and yellow) and ocher (chromium oxide) in an amount from 1% to 5%. The blue and yellow pigments have a phthalocyanine organic chemical base.

The binders were mixed with drinking water with a pH of 7.12.

A quartz sand (Ukhta, Russia) with a real density of 2611 kg/m^3, a size module equal to 1.7, and a bulk density of 1490 kg/m^3 was used as a filler.

The granulometric composition of the sand is given in Table 1.

For comparison, mineral additives such as silica fume, wollastonite and diatomite were used.

Due to the high-water demand of lime compositions in order to regulate the rheological, technological and functional properties of the finishing compositions, the following plasticizing additives were introduced into the mix recipe:

Kratasol that are salts of naphthalene sulfonic acids with a high content of high molecular weight fractions. S-3 (sulfonated poly condensates up to 82–84%; sodium sulfate is 8–10%; moisture no more than 10%). SP-3 is a mix of sodium salts of polymethylene naphthalene sulfonic acids of various molecular weights, industrial lignosulfonates, as well as industrial mix of sodium thiosulfate and sodium thiocyanate.

Table 1. Granulometric composition of the sand

Remaining name	Residue on sieves					Sieve pass 0.14
	2.5	1.25	0.63	0.315	0.14	
Partial	1	2.1	9.6	39.5	44	3.8
Complete	1	3.1	12.7	52.2	96.2	100

Hidetal P-4 (SKT-Standard, Novozybkov, Russia), Melment F 15 G and Melflux 1641 F (BASF, Berlin, Germany) polycarboxylate-based superplasticizers. The amount of the additive was 0.7–2% by weight of the binder.

To increase the water retention capacity, the additive Mecellose FMC 2094 (Lotte Fine Chemicals, Seoul, Korea) was also introduced into the formulation of the lime-sand composition in an amount of 0.1% by weight of the dry mix. To increase the strength characteristics, redispersible powders were also introduced into the formulation of the lime-sand composition: Pulver DM 1142P (Pulver, Turkey) Neolith 4400, Neolith 7200 (Neolith, Italy) in the amount of 1%, 0.8%, and 1.8% of the dry mix mass, respectively. They contain anticoagulants and anti-caking agents. Pulver DM1142P is a redispersible powder in the form of vinyl acetate, ethylene, and polyvinyl alcohol monomers.

When hardening in air, the lime mortars made with slaked lime give significant shrinkage. In this regard, the quartz sand of fraction 0.63–0.315 and 0.315–0.14 mm, in a ratio of 80:20 and bulk density of 1527 kg/m^3, was introduced into the lime mortar as a fine aggregate. In this case, the ratio of binder: sand (B:S) was 1:3 and 1:4.

The mix design was conducted based on the different ratio of the addition of the synthesized silicate hydrates to lime, as well as from the different water: lime ratios

For each series of tests, six specimens of each composition were made. Moreover, the standard deviation of the results did not exceed 5% for all results

The obtained filler's granulometric composition was assessed using an Analysette 22 automatic laser diffractometer (Fritsch, Germany). The determination of the specific surface area of powder materials was conducted by air permeability on a PSH-11 device (Stroydetal, Moscow, Russia). XRD patterns of the studied samples were obtained on an ARL X'TRA diffractometer (Thermo Electron, Waltham, MA, USA). SEM images were obtained using a G2 pro scanning electron microscope (Phenom-world BV,

Eindhoven, The Netherlands). Images of CSH crystals were also obtained using an optical microscope (Nikon, Tokyo, Japan). IR spectrometry was conducted using a 1000 FT-IR instrument (Varian Scimitar, Palo Alto, USA). Differential thermal analysis of the samples was conducted using a DTG-60H derivatograph (Shimadzu, Japan).

The real density of the starting materials was determined using a Le Chatelier instrument. The bulk density was determined by filling a vessel with a capacity of 1000 cm^3. The average density was determined as the ratio of the sample mass to its volume. The materials' sorption capacity was studied by keeping them in desiccators at different humidity levels and for various times. Their solubility determined the activity of the synthesized mineral additive in a 20% KOH solution. Determination of the amount of cohesive lime was determined following the Russian State Standard GOST 22688-2018 "Lime for building purposes". Test methods to determine the total content of active CaO + MgO in calcium lime using a titration method based on the neutralization reaction of hydrated lime HCl:

$$CaO + 2HCl + H_2O = CaCl_2 + 2H_2O$$

The investigation of finishing layers for strength in compression and bending was researched using a setup developed by the authors on specimens 10 × 10 × 10 mm and 10 × 10 × 50 mm in size, respectively (Figure 1).

The porosity was determined by the method of mercury porosimetry. Water absorption was determined according to the Russian standard GOST 5802-86 using 1 × 1 × 5 cm samples at the age of 28 days.

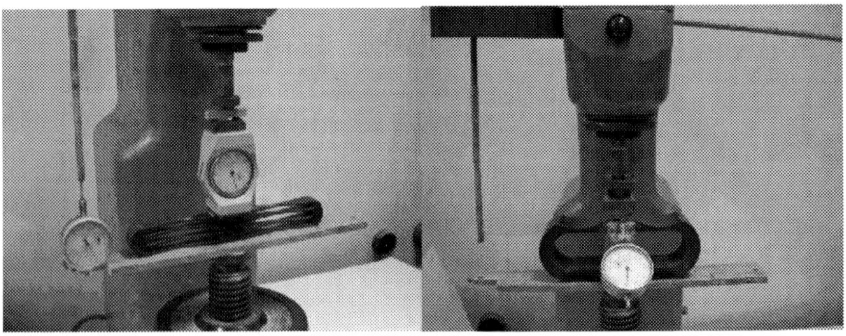

(a) (b)

Figure 1. Setup for determining the strength in bending (a) and compression (b).

The water resistance was evaluated by the softening coefficient, which is equal to the ratio of the compressive strength of the material saturated with water to the compressive strength of the dry material

$$K_s = \frac{R_s}{R_a} \tag{1}$$

The finishing mixes' water retention capacity was determined following the Russian State Standard GOST 5802-86.

The rheological properties of the compositions were evaluated in terms of plastic strength, which was determined using a CP-3 conical plastometer. During the experiment, the influence of the mix formulation on the change in the rheological properties of the formulations was investigated. The variable factors were the ratio of the components "lime: CSH filler", the conditions for the synthesis of the filler, the type and concentration of additives-plasticizers.

To assess the fracture toughness of the lime composites, the measurements of shrinkage deformations were conducted using an IZA-2 optical comparator.

Water resistance was evaluated by the softening coefficient, which is equal to the ratio of the compressive strength of a material saturated with water to the one of a dry material.

The water-holding capacity of the finishing compounds was determined in accordance with the Russian standard GOST 5802.

To assess the operational resistance of coatings based on lime-based dry mixes, frost resistance tests were conducted by alternating thawing and freezing of the finishing layer applied to the cement-sand substrate after 28 days of air-dry hardening. The assessment of the appearance of the coatings was conducted according to the Russian standard GOST 6992-68.

Results and Discussion

Synthesis of Calcium Silicate Hydrates

In developing the technology for the production of the fillers, the following factors were considered: a density of the liquid glass, an amount of addition of the precipitant and the rate of its introduction, and a mode of drying the precipitate and the time of its storage.

In the course of the research, the optimal content of the precipitating additive in the synthesis of the filler was revealed, which is 30-50% of the mass of water glass in the form of a 7.5-15% solution, depending on the density and modulus of the water glass.

From Table 2 it can be seen that with an increase in the content of the precipitating additive during the synthesis of the filler, the compressive strength of lime compositions with fillers increases. Thus, the compressive strength of a limestone with a filler synthesized from liquid glass with a density of $\rho = 1335$ kg/m^3 and a modulus of $M = 2.9$ in the presence of a 15% solution of $CaCl_2$ additive in an amount of 15-90% of the mass of water glass per dry residue, varies in the range of values 2.34-4.98 MPa, respectively, and with a filler synthesized from water glass in the presence of a 7.5% solution of $CaCl_2$ additive, the compressive strength is in the range of 2.42-5.07 MPa. The yield of fillers synthesized from water glass in the presence of $CaCl_2$ in the form of a 15% solution in an amount of 30% and 50% of the weight of water glass was 85%, and for fillers synthesized from water glass in the presence of $CaCl_2$ in the form of 7.5% solution in the amount of 30% and 50% - 100%. The compressive strength of limestone with a filler synthesized from water glass with a density of $\rho = 1663$ kg/m^3 and a modulus of $M = 1.53$ in the presence of a 15% solution of $CaCl_2$ additive in an amount of 15-90% of the weight of water glass is in the range 1.78-3.86 MPa. The output of the filler synthesized from liquid glass with a density of $\rho = 1663$ kg/m^3 and a modulus of $M = 1.53$, depending on the concentration of the $CaCl_2$ additive, is 36-95%. An increase in the content of a precipitating additive during the synthesis of CSH leads to an increase in the compressive strength of the limestone. However, the output of CSH is declining.

Table 2 shows that with an increase in the precipitate additive content during the filler synthesis, the lime compositions' compressive strengths with fillers increases.

It was found that the output of the filler synthesized from the water glass in the presence of $CaCl_2$ in the form of a 15% solution in an amount of 30 or 50% of the weight of liquid glass was 85%. This refers to the first set of specimens (liquid glass modulus 2.9). The output of the filler synthesized from the liquid glass in the presence of $CaCl_2$ in the form of 7.5% solution in the amount of 30 or 50% of the weight of liquid glass was 100%. After drying at a temperature of 105°C, the filler's real density was 2200 kg/m^3, and the bulk density was 1935 kg/m^3.

Table 2. Compressive strength of lime composites

Liquid glass module	Density of liquid glass, kg/m³	The amount of $CaCl_2$ precipitating additive based on the mass of liquid glass, %	Concentration of the solution of the additive-precipitant, %	Compressive strength of lime compositions in the presence of CSH, MPa	CSH output, %
2.9	1335	15	7.5	2.42	54
			15	2.34	55
		30	7.5	4.72	100
			15	4.60	85
		50	7.5	4.84	100
			15	4.76	85
		90	7.5	5.07	90
			15	4.98	67
1.53	1664	15	7.5	1.71	38
			15	1.78	36
		30	7.5	3.24	55
			15	3.28	53
		50	7.5	3.64	95
			15	3.72	88
		90	7.5	4.28	85
			15	3.86	0

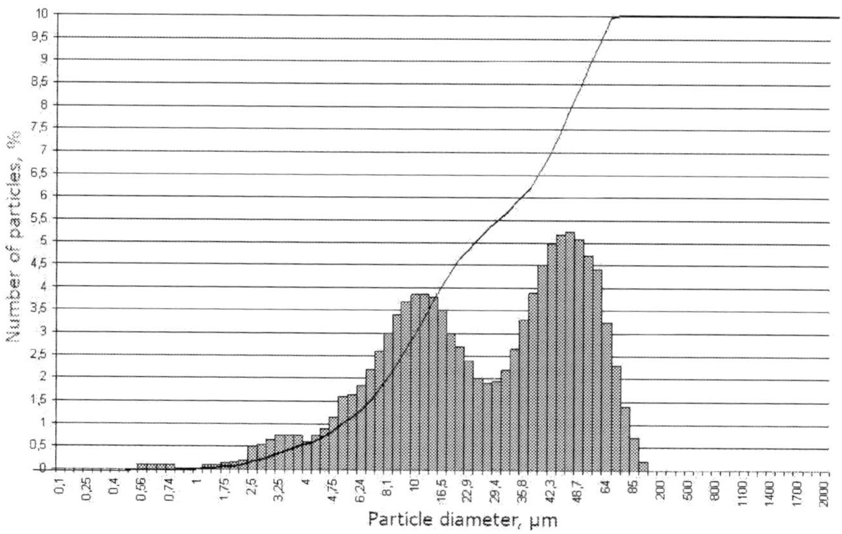

Figure 2. Granulometric composition of the fillers.

It was revealed that the particle size distribution of the silicate-containing filler has a similar polyfractional nature; the size distribution of the filler particles is bimodal (Figure 2).

The slow introduction of the precipitate solution was conducted within 10 min, and the fast one was within 1 min. With the slow introduction of the $CaCl_2$ precipitating additive, a small number of large particles were formed in the sediment. The granulometric data indicate that particles with sizes of 20–45 μm prevail, the contents of which was in the range 34–39%. The arithmetic mean value of the filler's particle size synthesized with the slow introduction of the $CaCl_2$ precipitating additive was 29.35 μm.

The use of more dilute solutions (7.5%) led to a slowdown in precipitation and the appearance of larger crystals. The content of particles with a size of 45–100 μm was 32.62%, crystals with a particle size of 100–200 μm appeared, and their content was 0.04%.

With an increase in the sediment's maturation time, crystal growth was observed (Figure 3). The content of particles with a size of 45–100 μm increased to 33.21%. The content of particles with a size of 100–200 μm increased by 1.02%.

An increase in temperature during deposition accelerates the formation of a crystal lattice and inhibits the formation of nucleation centers of crystallization, resulting in the formation of larger crystals (Figure 4)

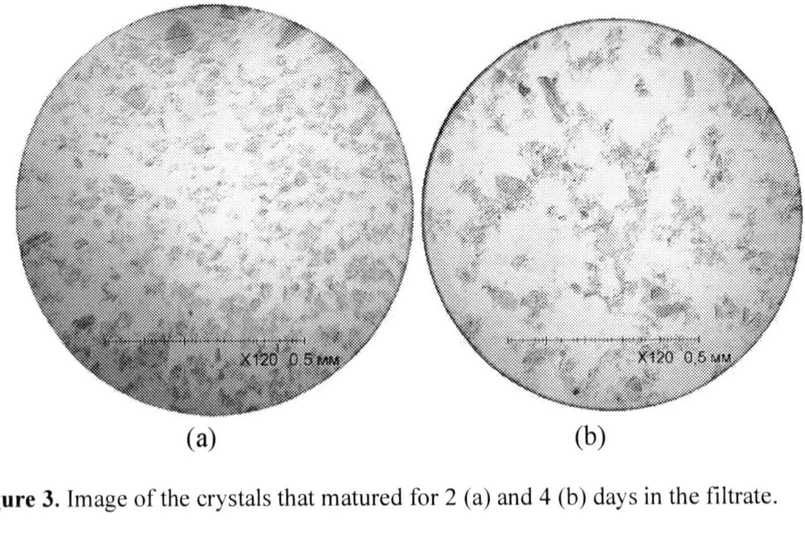

(a) (b)

Figure 3. Image of the crystals that matured for 2 (a) and 4 (b) days in the filtrate.

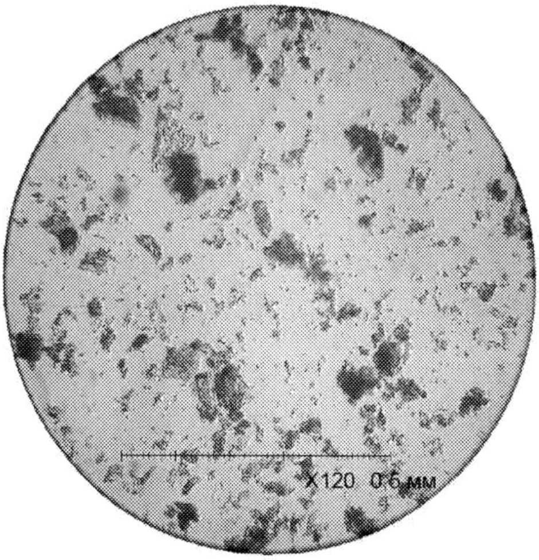

Figure 4. Image of CSH crystals as a result of synthesis from heated liquid glass with the introduction of 15% $CaCl_2$.

It was found that with the slow introduction of a precipitating additive (dropwise) in the form of a 15% solution, a small number of small particles are formed in the sediment. Granulometric data indicate that the content of particles with a size of 0.05-10 microns is 18.35%, particles with a size of 20-45 microns predominate, the content of which is 38.52%.

With the rapid introduction of a precipitating additive, the rate of formation of crystal nuclei is high and a finely dispersed precipitate is formed, consisting of a large number of small particles. It was revealed that the arithmetic mean of the size of the filler particles is 28, 41 microns, with 35% being particles with a size of 15.579 microns. Less than 5% are particles with a diameter of 2.954 microns, and with slow introduction - with a diameter of 4.279 microns.

In the course of the research, the pH of the solutions was determined. The pH of a 7.5-15% $CaCl_2$ additive solution is 7.12-7.3, respectively; A 50% solution of liquid glass with a density of $\rho = 1335$ kg/m^3 and a modulus of M = 2.9 has pH = 11.8, while a 50% solution of liquid glass with a density of 1663 kg/m^3 and a modulus of M = 1.5 pH = 13.6. The pH of the precipitate obtained as a result of the synthesis of solutions of water glass and a precipitating additive, depending on its content, as well as the modulus and density of the water glass, is shown in Table 3.

Table 3. Influence of sedimentation conditions on the pH of the precipitate

The amount of $CaCl_2$ precipitating additive	Liquid glass module	Density of liquid glass, kg/m^3	pH
30% of the mass of liquid glass	2.9	1335	11
50% of the mass of liquid glass			10.2
90% of the mass of liquid glass			9.6
30% of the mass of liquid glass	1.53	1663	12.6
50% of the mass of liquid glass			12.4
90% of the mass of liquid glass			12.2

Table 4. Compressive strength of lime composite

Filler synthesis conditions	Compressive strength, MPa
Precipitation in the presence of an additive of the $FeCl_3$ chromophore in the form of a 15% solution in an amount of 30% by weight of water glass	2.86
Precipitation in the presence of $CaCl_2$ additive in the form of a 15% solution in an amount of 30% of the mass of water glass	3.62
Precipitation in the presence of a precipitating additive $CaCl_2$ (1/7 of the mass of water glass) and an additive chromophore $FeCl_3$ (1/3 of the mass of water glass) in the form of a 15% solution	4.63
Precipitation in the presence of a precipitating additive $CaCl_2$ (1/2 part of the mass of water glass) and an additive chromophore $FeCl_3$ (1/2 part of the mass of water glass) in the form of a 15% solution	4.32
Precipitation in the presence of a precipitating additive $CaCl_2$ (1/3 of the mass of water glass) and an additive chromophore $FeCl_3$ (1/7 of the mass of water glass) in the form of a 15% solution	3.96

To obtain colored fillers, an additive-chromophore $FeCl_3$ was introduced, both together with the addition of $CaCl_2$, and separately. An analysis of the experimental data showed that the compressive strength of lime compositions with a filler synthesized from water glass in the presence of the combined introduction of $CaCl_2$ and $FeCl_3$ additives is 28% higher compared to compositions based on fillers synthesized in the presence of only $CaCl_2$ (Table 4).

The strength of lime composites with a composition of 1:0.3 at W/L = 0.9 with the use of colored fillers based on $FeCl_3$ varies in the range of 2.86-4.63 MPa, while when using uncolored fillers (control sample) it is - 3.62 MPa.

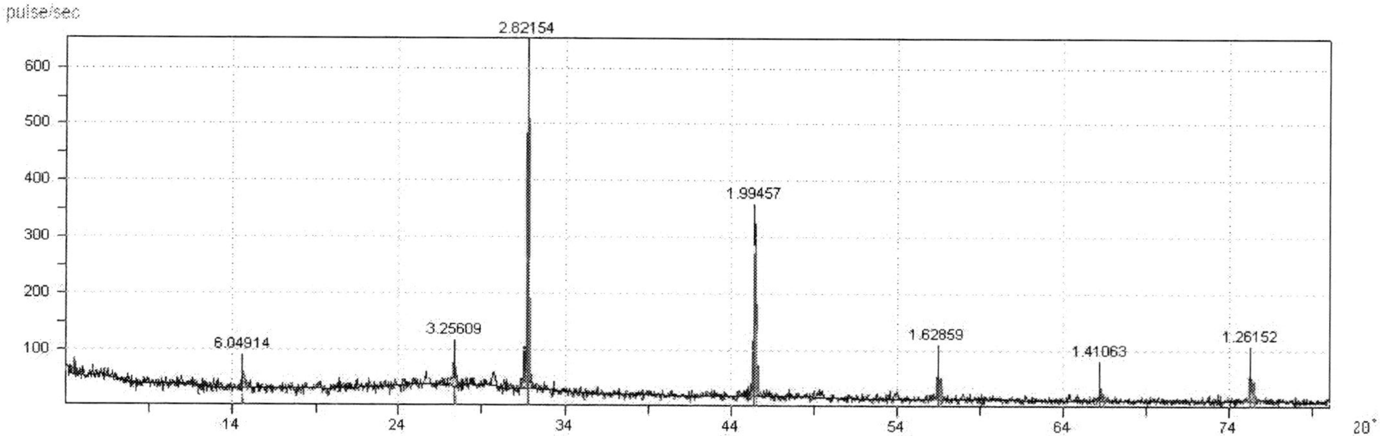

Figure 5. X-ray diffraction pattern of filler samples synthesized with the introduction of $CaCl_2$ additive in an amount of 50% of the mass of the water glass with a silicate modulus of 2.9.

In studying the new growth's qualitative composition, it was found that the samples' degree of crystallization is low. The X-ray diffraction pattern (Figure 5) of the filler samples contains diffraction lines (Å) of the following compounds: calcium silicate hydrates of the tobermorite group (10.13059; 3.58269; 3.25556; 3.2579; 2.82015; 2.4662; 1.29764; 1.2618); solid solution CSH (B) in the form of slightly crystallized gel (4.76541; 3.03952; 2.82163) solid solution C–S–H (II): (2.22058; 1.87721; 1.41032); hydrohalites: (3.85831; 1.99449; 1.62748).

The filler structure is represented by lamellar and acicular formations corresponding to calcium silicate hydrates (Figure 6).

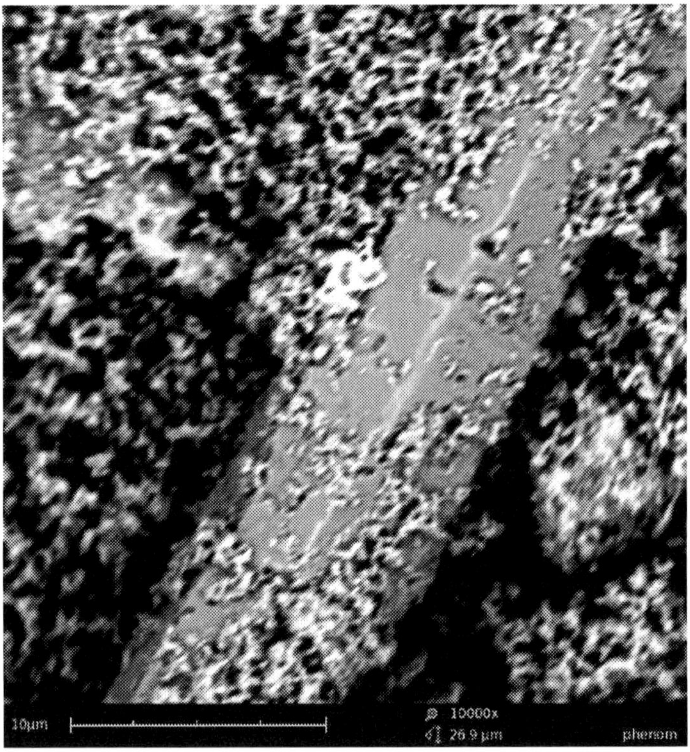

Figure 6. SEM image of the synthesized powder (x10000).

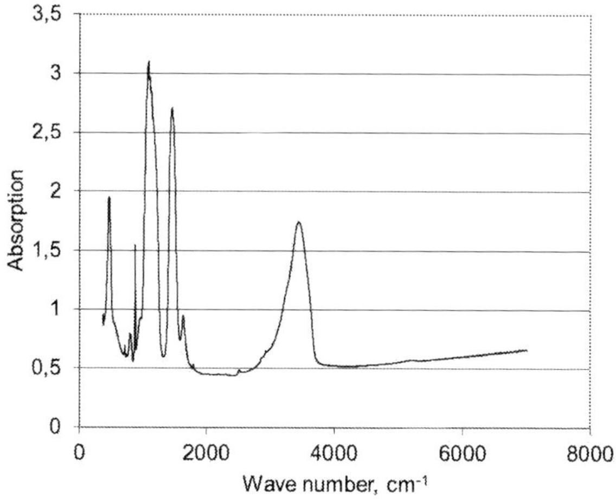

Figure 7. IR spectrum of the synthesized calcium silicate hydrates (CSH) filler sample.

Analysis of the IR spectrum of the obtained filler's sample was performed for additional assessment. Figure 7 shows clearly distinguished absorption bands in the regions of 850–1100 cm^{-1}, 550–750 cm^{-1}, 400–550 cm^{-1}, confirming the presence of calcium silicate hydrates in the synthesized material.

Studies have shown that the CSH-based filler has hydraulic activity.5 shows that the activity of the filler varies depending on the synthesis mode in the range of 178–289 mg/g.

The activity of the fillers depends on the drying temperature (Table 6). The most active is the filler dried after filtration at a temperature of 300°C. Compressive strength R_{com} of specimens at the age of 28 days of hardening in air-dry conditions, of the composition 10.3 (lime: CSH filler) by weight at a water: lime ratio equal to 0.7, when using the filler dried at temperature of 300°C, is R_{com} = 6.79 MPa. In addition, the composition with the use of the filler dried at a temperature of 105°C is 4.56 MPa (i.e., the increase in compressive strength is 50%).

In order to study the influence of the terms and conditions of storage of the filler on its activity, part of the filler was stored in conditions excluding the access of moisture, and remaining part was stored in the open air at a relative humidity of 70–75% and a temperature of 18–20°C. After storage of the filler for 10–40 days, specimens of the composition lime: CSH (filler) = 1:0.3 were molded with water: lime ratios of 0.7 and 0.9. The specimens hardened at a relative air humidity of 70% and a temperature of 18–20°C (Table 7). Specimens molded immediately after drying of the fillers were taken as control ones.

The research results shown in Table 7 indicate that during storage of the filler under conditions excluding moisture access, the activity of the filler practically does not change. Therefore, the value of the compressive strength of the samples molded at W/L = 0.65 on the filler immediately after drying is R_{com} = 4.56 MPa and molded on the filler after storage for 10–20 days is 4.56–4.50 MPa. Some fluctuations in compressive strength values are associated with statistical variability. After 20 days of storage in conditions that exclude moisture access, its activity decreases slightly. The decrease in the compressive strength of specimens molded at W/L = 0.7 on the filler after storage for 30–40 days is 8.3–19.7%, and at w/l = 0.9 is 10.5–24.6%.

The study of the hygroscopic properties of the fillers showed that they have a high sorption capacity. At a relative humidity of 72%, the sorption humidification after 10 days was 20%, and at a relative humidity of 100% it was 95%.

It was found that storage of the filler under conditions that exclude moisture access does not change its activity. After 20 days of storage under conditions excluding moisture access, the decrease in the activity of the filler was 5%, and after storage for 30–40 days it was 9–20%.

Table 5. Filler activity

Filler synthesis conditions	**Filler activity, mg/g**
Filler synthesized from liquid sodium glass with density ρ = 1335 kg/m^3 and modulus M = 2.9 in the presence of $CaCl_2$ precipitating additive in an amount of 30% of its mass	289
Filler synthesized from liquid sodium glass with density ρ = 1130 kg/m^3 and modulus M = 2.9 in the presence of $CaCl_2$ precipitating additive in an amount of 30% of its mass	215
Filler synthesized from liquid sodium glass with density ρ = 1335 kg/m^3 and modulus M = 2.9 in the presence of $CaCl_2$ precipitating additive in an amount of 50% of its mass	243
Filler synthesized from liquid sodium glass with density ρ = 1130 kg/m^3 and modulus M = 2.9 in the presence of $CaCl_2$ precipitating additive in an amount of 50% of its mass	225
Filler synthesized from liquid sodium glass with density ρ = 1663 kg/m^3 and modulus M = 1.53 in the presence of $CaCl_2$ precipitating additive in an amount of 30% of its mass	178
Filler synthesized from liquid sodium glass with density ρ = 1663 kg/m^3 and modulus M = 1.53 in the presence of $CaCl_2$ precipitating additive in an amount of 50% of its mass	285

Table 6. Strength of lime compositions

Filler drying temperature, °C	Water-lime ratio, W/L	Strength, MPa	
		when compressed	when flexion
105	0.7	4.56	1.73
105	0.9	3.62	1.59
200	0.7	6.15	1.84
200	0.9	4.88	1.73
300	0.7	6.79	2.12
300	0.9	5.69	1.98

Table 7. Effect of storage conditions on the filler activity

Water-lime ratio W/L	Compressive strength, MPa				
	Filler storage time, days				
	0	10	20	30	40
0.7	4.56	4.56/4.32	4.50/3.82	4.18/3.40	3.66/2.84
0.9	3.62	3.62/1.82	3.62/1.82	3.24/1.76	2.73/1.37

Note - above the line are the values of the compressive strength during storage of the filler in conditions excluding the access of moisture, below the line - during storage in air.

The generalized analysis of the results of experimental data made it possible to establish the optimal ratio of components, select the type, amount, and concentration of the solution of the precipitant additive involved in the synthesis of the filler, and also establish the optimal density and modulus of the liquid glass. The greatest advantages in terms of strength characteristics are possessed by compositions with the filler obtained by synthesis from the liquid glass, with a density of 1335–1663 kg/m^3 and a silicate module of 1.53–2.9, in the presence of a precipitating additive $CaCl_2$ in an amount of 30–50% of the weight of the liquid glass in the form of a 7.5–15% solution, and dried at a temperature of 105°C.

Development of the Modified Lime Compositions

The influence of the water–lime ratio (W/L), the content of modifying additives, the CSH filler, and the technology of its production on the properties of the finishing compositions was evaluated. It was revealed that the introduction of filler based on CSH into the lime composition changes the character of structure formation.

A thermodynamic analysis of possible reactions was carried out during the interaction of the mix components (lime, filler based on calcium silicate hydrates, and mixing water) in accordance with the second law of thermodynamics. The law establishes a connection between the thermal effect of a chemically irreversible process and the work of the corresponding irreversible process and is determined by the Gibbs-Helmholtz equation:

$$\Delta G = \Delta H_p + T\frac{\partial \Delta G}{\partial T} \tag{2}$$

where ΔG is the Gibbs energy;

ΔHp - enthalpy of the process;

T - temperature, K.

Taking into account that the structure formation of the system under study proceeds at a temperature (293K) close to the standard temperature (298 K), the thermal effect of the reaction and the change in the Gibbs energy were determined only at the standard state and were calculated as the difference between the sums of the corresponding indicators of the reaction products and the initial substances (Table 8).

The calculation results indicate the probability of all considered reactions proceeding in the forward direction (negative values of ΔH^0_{298}). The large numerical value ΔG^0_{298} found for the formation reaction makes it possible with sufficient probability to speak about the possibility of the reaction proceeding at standard temperature (25°C) and at other temperatures.

The hardening of solutions on slaked lime is due to the occurrence of two processes: Crystallization of $Ca(OH)_2$ during drying of solutions and carbonization of hydroxide. This process takes place primarily in the surface layers. Carbonization of deep layers is long-lasting, because, firstly, the amount of CO_2 in the atmosphere is only 0.04%, and, secondly, the resulting $CaCO_3$ film has low permeability. Therefore, in the central part of well-compacted solutions, a significant amount of $Ca(OH)_2$ remains for a long time. Evaporation of water from the solution also helps to increase the strength. The formation of $CaCO_3$ leads to an increase in the strength and water resistance of products. If active additives are used as a filler, along with the formation of carbonates, calcium silicate hydrates may also appear, which increase the strength of solutions. The formation of a significant amount of silicate hydrates that improve the adhesion of the binder to the aggregate.

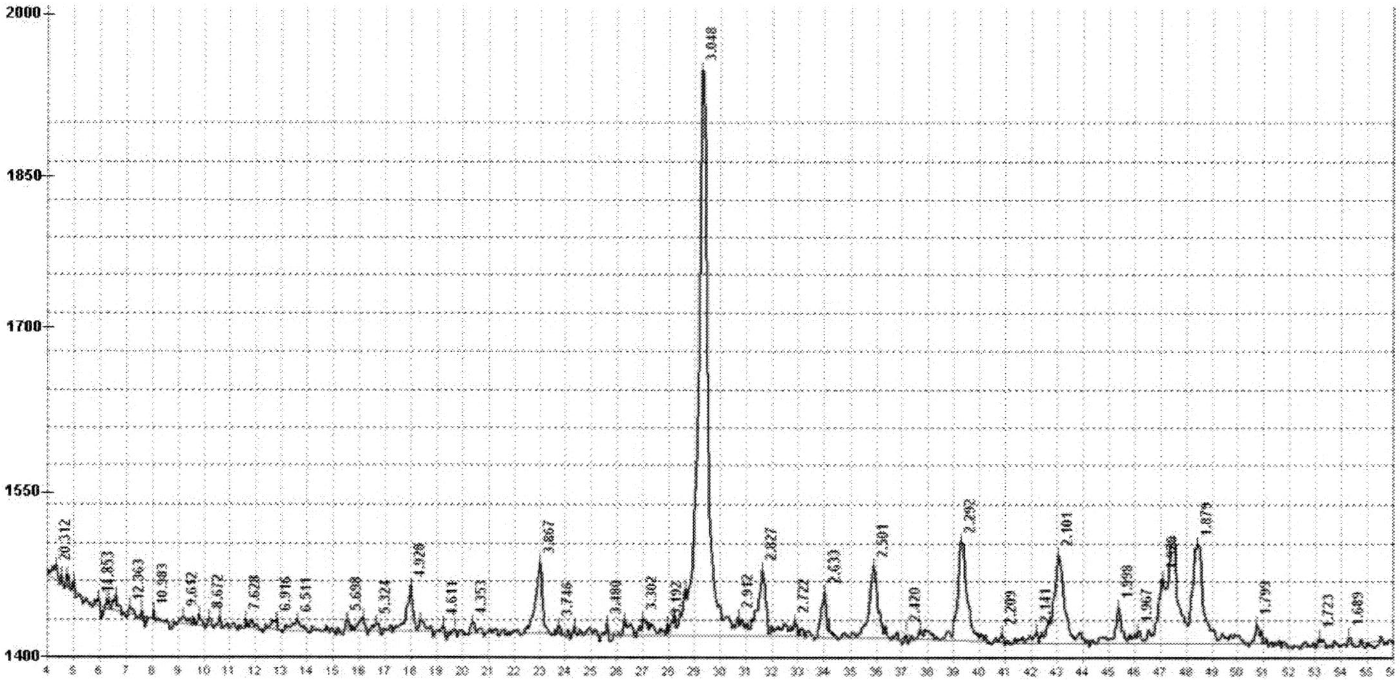

Figure 8. X-ray diffraction pattern of samples 03-A of a lime composite with a filler synthesized with the addition of $CaCl_2$ in an amount of 30% of the mass of the water glass with a silicate modulus of 2.9.

Table 8. Thermodynamic parameters of reactions in the interaction of the components of the mortar mix

Reaction number	Compound formula	Heat of formation ΔH⁰298, kJ/mol	Gibbs energy ΔG⁰298, kJ/mol	Reactions
1.	$Ca(OH)_2$	-63,3	-55,7	$CaO + H_2O = Ca(OH)_2$
2.	$CaCO_3$	-114,7	-74,6	$Ca(OH)_2 + H_2O = CaCO_3 + H_2O$
3.	$3CaO \cdot 2SiO_2 \cdot 3H_2O$	-2757	-2439,4	$5CaO \cdot 6SiO_2 \cdot 5H_2O + Ca(OH)_2 = 3(3CaO \cdot 2SiO_2 \cdot 3H_2O)$

In the study of solid-phase reactions occurring in the process of structure formation of lime compositions (Figure 8) it was revealed that in the samples of the lime composite with the synthesized filler there were diffraction lines (Å) of the following compounds: afwillites: 3.30452; 2.72286; portlandites:4.928; 2.633; 2.501;1.967;1.799; calcites:3.747; 3.048; 3.501; 2.912; 2.633; 2.42; 2.141;1.998; hillebrandites: 1.9242; 1.8730; okenites: 7.67421; tobermorites: 3.192; 2.82723; calcium silicate hydroxides: 3.83385; 2.09187.

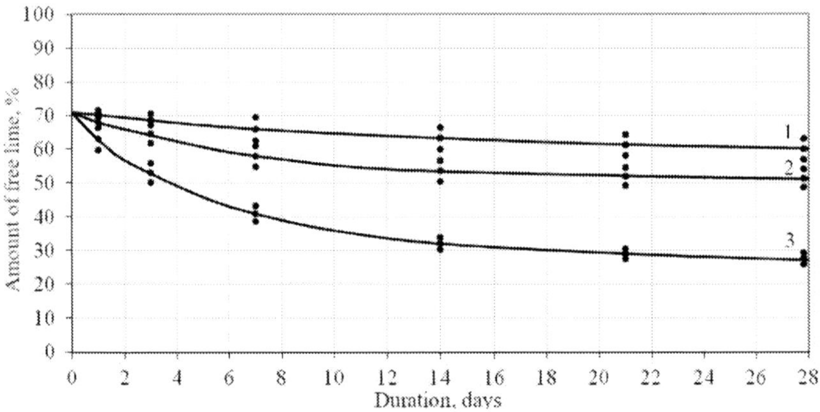

Figure 9. Decrease in the amount of free lime over time: 1— control composition; 2— composition L: wollastonite = 1:0.3, W/L = 0.9; 3— composition L: filler based on CSH = 1: 0.3, W/L = 0.9.

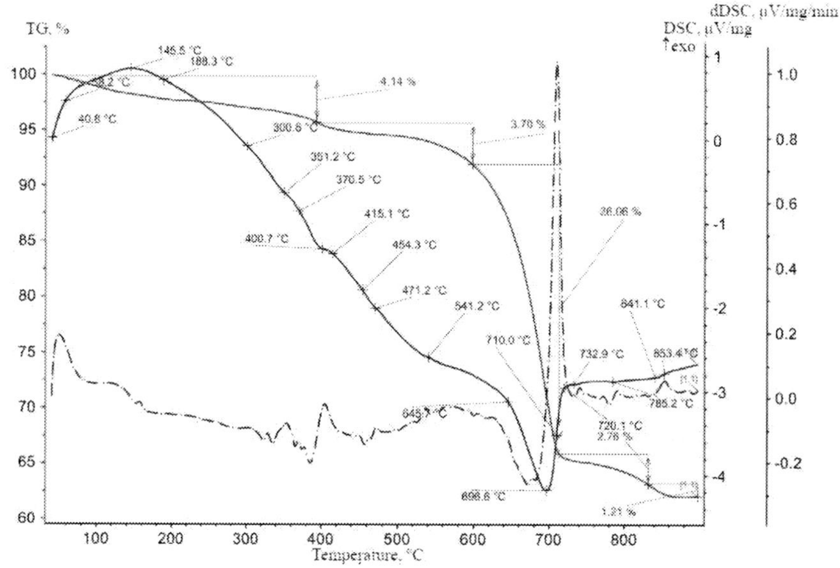

(a)

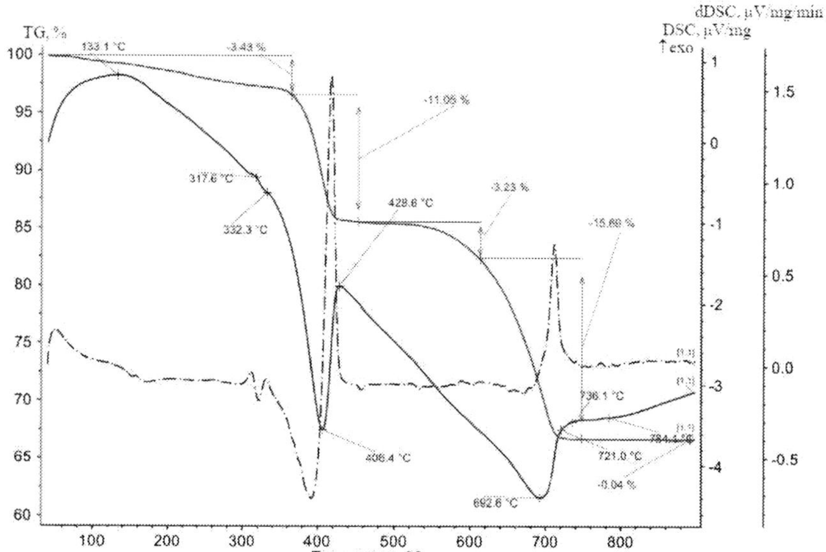

(b)

Figure 10. DTGA patterns of lime composites: (a) with a filler based on CSH; (b) – control.

Table 9. Strength of the lime composite depending on the filler content

Composition	Water-lime ratio W/L	Ultimate strength, MPa, at the age of 28 days	
		When compressed	When flexural
control	0,7	1.62	0.42
	0,9	1.56	0.40
lime: filler CSH = 1:0.05	0.7	3.96	1.38
	0.9	3.24	1.26
lime: filler CSH = 1:0.15	0,7	4.56	1.73
	0,9	3.82	1.63
lime: filler CSH = 1:0.3	0,7	4.56	1.73
	0.9	3.62	1.59
lime: filler CSH = 1:0.5	0.9	4.31	1.80
	1.0	3.82	1.80
lime: diatomite = 1:0.3	0.9	2.06	0.72
lime: filler CSH: diatomite = 1:0.1:0.2	0.9	3.46	1.06
lime: filler CSH: diatomite = 1:0.2:0.1	0.9	3.12	0.96
lime: wollastonite = 1:0.3	0.7	2.06	1.06
	0.9	1.86	0.96
lime: wollasto-nite = 1:0.15	0.7	1.93	0.72
	0.9	1.74	0.71
lime: micro silica = 1:0.3	0.7	2.12	1.34
	0.9	2.09	1.18
lime: micro silica = 1:0.15	0.7	2.24	1.14
	0.9	2.06	1.05

An increase in the number of peaks in the X-ray diffraction pattern of samples of a lime composite with filler based on CSH suggests that this filler was highly active and indicates a chemical interaction with lime. The studies conducted show that the CSH filler has a hydraulic activity of 178–289 mg/g depending on the synthesis mode. It was found that the amount of free lime in the control samples after 28 days of air-dry hardening was 60%, and in the samples with the CSH filler of the composition lime: CSH = 1:0.3, w/L = 0.9– 27%, which indicates about the chemical interaction of calcium silicate hydrates with lime (Figure 9).

The data obtained were further confirmed during the differential thermal analysis (Figure 10).

It was found that in the samples of a lime composite with CSH filler, along with the decomposition of calcium oxide and calcium carbonate, dehydration of tobermorite (454.2°C), afwillite (471.2°C), hillebrandite (541.2°C), okenite

(645.7°C) was observed, whose products are calcium silicates and modifications, wollastonite and cristobalite (Figure 10a). The total weight loss in the lime composite samples with the filler in the temperature range 400–700 °C was 26%, and in the control, sample was only 5% (Figure 10b). A significant loss of water by weight in the samples of the lime composite with the filler suggests that a large number of highly basic calcium silicate hydrates were formed as a result of the chemical reaction of the filler and the lime.

The obtained XRD and DTA data indicate the formation of a stronger structure of the lime composite with CSH filler. Table 9 shows the compressive strength data of the lime composite depending on the filler content. Compositions with wollastonite M200, diatomite, and micro silica were also studied for comparison.

The highest values of strength indicators for modified compositions were ratios of lime: filler = 1:0.3 at w/L = 0.7 and lime: filler = 1:0.5 at w/L = 0.9. The use of silica fume in the amount of 15% and 30% of the mass of lime leads to an increase in strength by 30-35%. The introduction of wollastonite into the formulation leads to an increase in compressive strength by 15-20%. The introduction of diatomite into the lime composition in an amount of 30% of the lime mass leads to an increase in strength by 32%. The joint introduction of diatomite with CSH leads to an increase in the strength of limestone by 100-122%.

An increase in the dispersion of the filler leads to an increase in its chemical activity and a regular increase in the values of the compressive strength (Table 10).

Table 10. Compressive strength of the lime composite depending on its specific surface area

Composition	Water-lime ratio W/L	Specific surface area S = 5896 cm^2/g	Specific surface area S = 7876 cm^2/g	Specific surface area S = 14223 cm^2/g
lime: filler CSH=1:0.3	0.7	3.64	4.56	6.

With an increase in the fineness of grinding of the filler based on CSH, an increase in the numerical values of the compressive strength of the limestone is observed. The highest strength is possessed by samples based on compositions with the use of a filler with a specific surface area S = 14223 cm^2/g. The compressive strength at the age of 28 days of hardening is R_{comp} = 6.2 MPa.

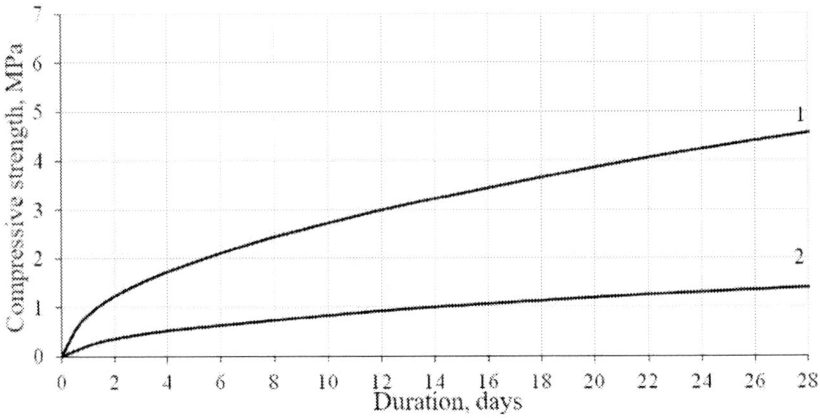

Figure 11. Kinetics of lime paste compressive strength gain: 1— with filler based on CSH; 2—- control.

Figure 11 shows the compressive strength values of lime paste during hardening. So, in 2 days, the compressive strength of lime paste with the CSH addition was 1.22 MPa, and for the reference composition was 0.36 MPa.

Based on the results shown in Figure 11, the models of the kinetics of hardening of the lime composites are proposed:

$$R = 0.86t^{1/2} \quad (3)$$

$$R = 0.26t^{1/2} \quad (4)$$

where R is a compressive strength of the lime composite; t is a hardening duration.

Equation (3) describes the kinetics of the strength gain of limestone with a filler based on CSH, and equation (4) – the control sample.

In order to regulate the rheological, technological, and functional properties of the finishing compositions, various superplasticizers were introduced into the mix formulation. The amount of the additive was 0.7–2% by weight of the binder. The use of superplasticizers S-3 or SP-3 gives the maximum plasticizing effect of the finishing lime composition. In this case, the water-reducing effect is 1.6 (Table 11) introduction of the CSH filler to the lime increases the water-reducing effect to 1.8. The water-reducing effect is a reduction in water consumption while maintaining equal flowability.

Table 11. Values of the water-reducing effect depending on the composition of the mixes

Name of the plasticizer	Content additives, % by weight of the binder	Binder composition	
		Lime	Lime: CSH = 1: 0.3
Kratasol (TU 5745-333-05800142)	0.4	1.3	1.6
Kratasol PL (TU 5745-333-05800142)	0.7	1.4	1.7
Kratasol PFM (TU 5745-333-05800142)	0.8	1.5	1.7
Hidetal P-4 (TU 5745-005-57330160-05)	0.9	1.4	1.7
S-3 (TU 6-36-0204229-625)	0.7	1.6	1.8
SP-3 (TU 5730-004-97474489-2007)	0.7	1.6	1.8
Melment^{R} F15G	0.4	-	-
Melflux^{R} 1641F	0.2	-	-
Melment^{R} F15G	2	1.5	1.7
Melflux^{R} 1641F	1.5	1.5	1.8

The research results showed that the plasticizing effect, depending on the type of additive, persists for 1–2 h. The plasticizing effect was not observed at low dosages of the additives Melment^{R} F15G (0.2–0.4%) and Melment^{R}1641F (0.2%). At the maximum dosage recommended by additive manufacturers, the water-reducing coefficient in CSH formulations was 1.7–1.8.

Table 12. Water retention capacity of the lime composites with the CSH filler

Composition	Water retention capacity, %
lime: filler (L: F) =1:0.3, Water-lime ratio W/L=0.9	94
lime: filler (L: F) =1:0.5, Water-lime ratio W/L=0.9	96
lime: filler (L: F) =1:0.3, Water-lime ratio W/L=0.9, S-3	95
lime: filler (L: F) =1:0.3, Water-lime ratio W/L=0.5, S-3	99

Table 13. Water retention capacity of the lime composites with the addition of a plasticizer

Composition	Water retention capacity, %
Water-binder ratio W/B = 1.8, binder: sand B: S = 1: 4	93.6
Water-binder ratio W/B = 1.8, binder: sand B: S = 1: 4, Mecellose FMC 2094	97.9
Water-binder ratio W/B = 1.8, binder: sand B: S = 1: 3	95.8

Table 13. (Continued)

Composition	Water retention capacity, %
Water-binder ratio W/B = 1.8, binder: sand B: S = 1: 4, S-3	94.5
Water-binder ratio W/B = 1.8, binder: sand B: S = 1: 3, S-3	98.6
Water-binder ratio W/B = 1.0, binder: sand B: S = 1: 4, S-3	98
Water-binder ratio W/B = 1.0, binder: sand B: S = 1: 3, S-3	98.3
Water-binder ratio W/B = 1.0, binder: sand B: S = 1: 4, S-3, Pulver DM 1142	98.4
Water-binder ratio W/B = 1.0, binder: sand B: S = 1: 4, S-3, Neolith 7200	98.6
Water-binder ratio W/B = 1.0, binder: sand B: S = 1: 3, S-3, Neolith 7200	98.7

The filler's introduction into the lime composition increased the water retention capacity up to 96%. The superplasticizer's introduction into the lime composition with the CSH filler increased the water retention capacity to 99% (Table 12, 13).

It was found that the use of the CSH fillers in the lime compositions, obtained with the rapid introduction of $CaCl_2$ additives into the liquid glass during the filler synthesis, accelerates the plastic strength gain of the lime compositions with this filler (Figure 12, curve 1). Thus, the plastic strength of the lime composition when using a filler obtained with the rapid introduction of $CaCl_2$ additive, in the form of a solution of 15% concentration at the age of 4.5 h, is $\tau = 0.025$ MPa, and the filler synthesized with the slow introduction of a precipitating additive is $\tau = 0.017$ MPa (Figure 12, curve 2). The filler, obtained by holding the filtrate for three days, slows down the development of plastic strength (Figure 12, curve 3).

It was found that the introduction of the S-3 superplasticizer leads to a slowdown in the gain of plastic strength (Figure 13, curve 5). A decrease in the amount of mixing water (w/b = 1.1), considering the water-reducing coefficient, naturally contributes to a faster gain of plastic strength in compositions with superplasticizers (Figure 13, curves 1, 2, 3). Thus, the plastic strength of the composition with the addition of S-3 at the time of 1 h of hardening is $\tau = 21.7$ kPa, for the 03C composition, $\tau = 5.2$ kPa. When the Kratasol additive is introduced into the formulation at the first stage of hardening (up to 1 h 20 min), there is a slight slowdown in the set of plastic strength compared to compositions containing the additives S-3 and SP-3;

however, subsequently the plastic strength of the composition with the addition of Kratasol is higher.

When evaluating the structure, it was found that lime compositions form coatings characterized by high porosity and a significant volume of open pores (Table 14).

Table 14. Characterization of the lime paste porous structure

Composition	Porosity, %	Open pore volume, %	Closed pore volume, %	Water absorption by weight, %
Controlling	60	50	10	44
Lime: filler (L: F) =1:0.3, Water-lime ratio W/L=0.9	56	43	13	38

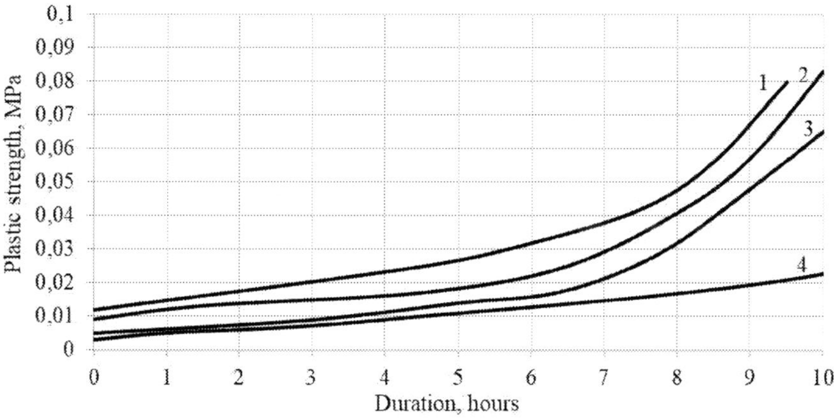

Figure 12. Change in the plastic strength of the specimen depending on the filler synthesis conditions: (1) the filler was synthesized with the rapid introduction of $CaCl_2$ additive in an amount of 90% of the water glass mass, the specific surface of the filler is 5876 cm^2/g; (2) the filler was synthesized with the slow introduction of $CaCl_2$ additive in an amount of 90% of the water glass mass, the specific surface of the filler is 4669 cm^2/g; (3) the filler was synthesized with the rapid introduction of $CaCl_2$ additive in an amount of 90% of the weight of the liquid glass, and kept for three days, the specific surface of the filler is 4439 cm^2/g; 4 - lime paste.

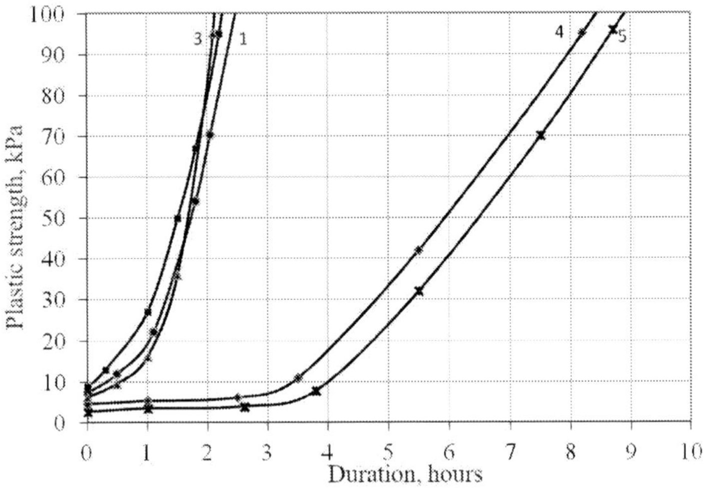

Figure 13. Change in the plastic strength of the lime specimens: 1 - W/B = 1.0, plasticizer S-3; 2 - W/B = 1.0, SP-3 plasticizer; 3 - W/B = 1.1, plasticizer Kratasol PFM; 4 - W/B = 1.8, (control); 5 – W/B = 1.8, plasticizer S-3.

The introduction of the CSH filler into the lime formulation leads to a decrease in porosity and an increase in the volume of closed pores; as a result of this, water absorption of the lime pastes decreases.

Table 15. Water resistance of finishing coatings

Composition of the composite	Additive	Softening coefficient
Lime: sand = 1: 4, W/B = 1.8	-	0.42
Lime: CSH: sand = 1: 0.3: 4, W/B = 1.8	-	0.57
Lime: sand = 1: 3, W/B = 1.8	-	0.45
Lime: CSH: sand = 1: 0.3: 3, W/B = 1.8	-	0.61
Lime: CSH: sand = 1: 0.3: 4, W/B = 1.0	S-3	0.68
Lime: CSH: sand = 1: 0.3: 3, W/B = 1.0	S-3	0.69
Lime: CSH: sand = 1: 0.3: 4, W/B = 1.0	SP-3	0.58
Lime: CSH: sand = 1: 0.3: 4, W/B = 1.0	Hidetal P- 4	0.63
Lime: CSH: sand = 1: 0.3: 4, W/B = 1.0	Kratasol PFM	0.71
Lime: CSH: sand = 1: 0.3: 4, W/B = 1.0	S-3, Pulver DM 1142 P	0.72
Lime: CSH: sand = 1: 0.3: 4, W/B = 1.0	S-3, Neolith 7200	0.73

Low values of open porosity and high values of strength of the developed composites made it possible to achieve high softening coefficients (Table 15).

So, the coefficient of softening of control samples makes $K_{soft} = 0.42$, in samples based on filler CSH – 0.54. Samples with Kratasol's PFM additive have higher value of coefficient of a softening (K_{soft} = 0.57) and S-3 (K_{soft} = 0.63). This is explained by existence in composition of an additive Kratasol PFM the hydrophobic component, and also by creation of more dense structure owing to reduction of an amount of water of a mixing. This is evidenced by the data on the kinetics of water absorption of the lime composites.

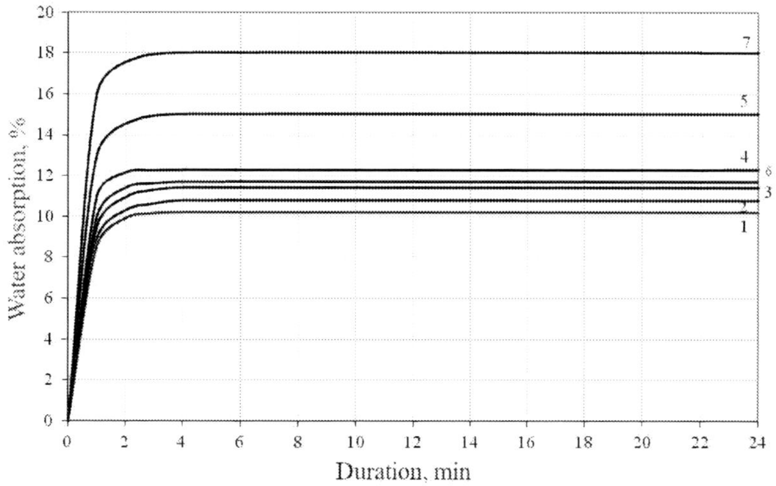

Figure 14. Water absorption of the lime composites: 1- composition W/B = 1.1, B:S = 1: 4, Kratasol PFM; 2 - composition W/B = 1.0, B: S = 1: 4, S-3; 3 - composition W/B = 1.1, B:S = 1:4, Hidetal P-4; 4 - composition W/B = 1.0, B: S = 1: 4, SP-3; 5 - composition W/B = 1.8, B:S = 1:4; 6-composition W/B = 1.0, B:S = 1: 3, S-3; 7- composition W/B = 1.8, B:S = 1:3.

It was found that, during the initial 1–4 min, all samples showed intense water absorption, and subsequently the water absorption indicators stabilized. Water absorption of the composition "water/binder = 1.8, binder: sand = 1: 4" after 24 h was 15% by weight (Figure 14, curve 5), and the composition "water/binder = 1.8, binder: sand = 1:3" was 18%. When a superplasticizer is introduced into the mix, a decrease in water absorption is observed. A lower value of the water absorption index, which is 10.2% at 24 h water absorption (Figure 14, curve 1), is observed in the samples prepared with the use of the Kratasol additive.

The numerical values of the water absorption coefficient, which for compositions with Kratasol and S-3+Pulver DM 1142P additives, respectively, were 0.49 and 0.39 $kg/(m^2 \cdot h^{0.5})$, indicate that coatings in accordance with DIN 52617 are water-repellent and hydrophobic (less than 0.5 kg/($m^2 \cdot h^{0.5}$) (Table 16).

Table 16. Water absorption during capillary suction

Compound	Additive	Water absorption during capillary suction	
		kg/(m^2 $h^{0,5}$)	kg/m^2
Lime: sand = 1:4, W/B = 1.76	-	1.1	4.8
Lime: CSH: sand = 1: 0.3: 4, W/B = 1.0	-	0.67	3.4
Lime: CSH: sand = 1: 0.3: 4, W/B = 0.94	S-3	0.56	2.8
Lime: CSH: sand = 1: 0.3: 4, W/B = 1.1	Kratasol PFM	0.49	2.4
Lime: CSH: sand = 1: 0.3: 4, W/B = 1.0	S-3, Pulver DM 1142P	0.39	1.9
Lime: CSH: sand = 1: 0.3: 3, W/B = 1.0	S-3, Neolith 7200	0.41	2.0

In the German standard DIN 18550 (part 2), it is noted that durability and resistance to external influences, as well as high fracture toughness, are provided when the plaster has a compressive strength in the range of values from 2 to 5 MPa. Mortars with such strength characteristics are able to adapt to small deformations and resist cracking. To increase the strength characteristics of coatings based on dry building mixtures (DBM), redispersible powders were introduced into the formulation: Pulver DM 1142P and Neolith 7200 in an amount of 1% and 0.8%, respectively, by weight of dry substances in accordance with the recommendations of the manufacturer FAR SPA (Italy).

To assess the fracture toughness of lime composites, measurements of shrinkage deformations were conducted using an IZA-2 optical comparator. The results of assessing the shrinkage deformations of the finishing compounds are shown in Figure 15.

Shrinkage deformations of specimens with ratios "lime: sand = 1:3 and water/binder = 1.8" after three months were 2.12 mm/m (Figure 15, curve 1). When the CSH filler is introduced into the lime composition, a decrease in shrinkage is observed, which is for the composition "lime: CSH = 1:0.3, binder: sand = 1:3 and water: binder = 1.8" after three months was 1.16 mm/m (Figure 15, curve 2). An increase in the amount of sand in the mix helps to reduce shrinkage. Shrinkage deformations of specimens with ratios "lime: CSH = 1:0.3, binder: sand = 1: 4 and water/binder = 1.8" were 0.76 mm/m (Figure 15, curve 3). The introduction of superplasticizer S-3 and redispersible powders such as Neolith 7200 and Pulver DM 1142P into the composition of the DPM also leads to a decrease in shrinkage deformations. Shrinkage value after three months of hardening for compositions "lime: CSH = 1:0.3, binder: sand = 1:3 and water/binder = 1.0+S-3+Neolith 7200" and "lime: CSH = 1:0.3, binder: sand = 1: 4 and water/binder = 1.0+S-3+Pulver DM 1142P" was 0.34 and 0.26 mm/m, respectively (Figure 15, curves 4 and 5).

Table 17. Crack resistance coefficient of lime coatings

Composition of the composite	**Crack resistance coefficient $K_{cr} = R_{flex}/R_{com}$**
Lime: sand (L:S) = 1:3, water-binding ratio W/B = 1.8	0.26
Lime: sand (L:S) = 1:4, water-binding ratio W/B = 1.8	0.28
Binder (lime: CSH = 1: 0.3): sand B: S = 1:3, water-binder ratio W/B = 1.8	0.35
Binder (lime: CSH = 1: 0.3): sand B:S =1: 4, water-binder ratio W/B = 1.8	0.37
Binder (lime: CSH = 1:0.3): sand B:S= 1:3, water-binder ratio W/B = 1.0, S-3, Neolith 7200	0.51
Binder (lime: CSH = 1:0.3): sand B:S=1:4, water-binder ratio W/B = 1.0, S-3, Pulver DM 1142P	0.49

In addition, the crack resistance of the coatings was assessed using the crack resistance coefficient, which was determined as the ratio of the flexural strength to the compressive strength (the higher it is, the higher the crack resistance). Table 17 presents data on the R_{flex}/R_{com} ratio. The samples were evaluated at the age of 28 days.

The crack resistance coefficient of lime compositions is 0.26-0.28, depending on the lime: sand ratio. The introduction of a filler based on CSH into a dry mixture increases the R_{flex}/R_{com} ratio by 32-35%. So, the coefficient of crack resistance of specimens of composition B (L: CSH = 1: 0.3): S = 1:3, W/B =1.8 is equal to 0.35, and for specimens of composition B (L: CSH = 1: 0.3): S = 1: 4, W / B = 1.8 - 0.37. With the introduction of redispersible powders, such as Neolith 7200 and Pulver DM 1142P in the presence of the S-3 plasticizer, considering the water-reducing coefficient in the finishing composition, the R_{flex}/R_{com} ratio increases by 75-96%. Crack resistance coefficient of samples of composition B (L: CSH = 1:0.3): S = 1:3; W/ B = 1, C-3, Neolith 7200 is 0.51, and samples of composition B (L: CSH = 1:0.3): S = 1:4; W/B = 1.0, S-3, Pulver DM 1142P - 0.49.

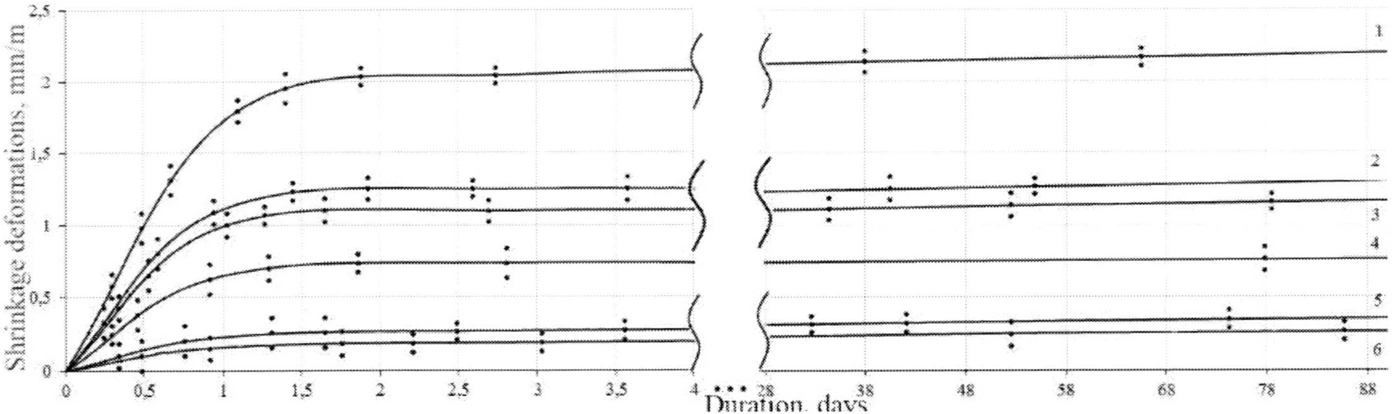

Figure 15. Shrinkage kinetics of finishing mixes:1 - composition L: S = 1: 3, W/B = 1.8; 2 - composition L: S = 1:4, W/B = 1.8; 3 - composition B (L:CSH = 1: 0.3):S = 1: 3, W/B = 1.8; 4 - composition B (L:CSH = 1:0.3): S = 1: 4, W/B = 1.8; 5 - composition B (L:CSH = 1: 0.3):S = 1: 3, W/B = 1.0, S-3, Neolith 7200; 6 - composition B (L:CSH = 1: 0.3):S = 1: 4, W/B = 1.0, S-3, Pulver DM 1142P.

To assess the operational resistance of coatings based on lime dry mixes, frost resistance tests were conducted by alternating thawing and freezing of the finishing layer applied to the cement-sand base after 28 days of air-dry hardening. After each cycle, a visual inspection of the surface was conducted in order to detect cracks, crumbling, and peeling of the material. The assessment of the appearance of coatings was conducted in accordance with GOST 6992-68 "Paintwork coatings. Method for determination of coating stability in atmospheric conditions". For "failure" was taken the condition of the coating, assessed by III.4 points. This condition is characterized by a loss of gloss up to 50% with significant discoloration, whiteness, bronzing and dirt retention, cracking up to 25% of the surface, the presence of a rash up to 25% and blisters up to 5% of the surface.

Table 18. Assessment of the decorative and protective properties coatings

Composition	The number of cycles	Points
Binder (lime: CSH = 1:0.3): sand, B: S = 1: 3, water-binding ratio W/B = 1.0, S-3, Neolith 7200 (control)	before testing	V. 8
	6	V. 8
	12	V. 8
	18	V. 8
	25	V. 7
	31	V. 7
	37	V. 6
	43	V. 5
	50	V. 5
Binder (lime: CSH = 1:0.3): sand, B: S = 1: 3, water-binding ratio W/B = 1.0, C-3, Neolith 7200, yellow SV pigment in an amount of 5%	before testing	V. 8
	6	V. 8
	12	V. 8
	18	V. 8
	25	V. 7
	31	V. 6
	37	V. 6
	43	V. 5
	50	V. 5
Binder (lime: CSH = 1:0.3): sand, B:S = 1: 3, water-binding ratio W/B = 1.0, S-3, Neolith 7200, FTZ blue pigment in an amount of 5%	before testing	V. 8
	6	V. 8
	12	V. 8
	18	V. 7
	25	IV. 5
	31	IV. 4
	37	III. 3
	43	III. 3
	50	III. 3

It was found that colored samples with a yellow pigment, along with control samples, withstood 50 test cycles. The condition of the coating after 50 test cycles was rated V.5 points. Table 18 shows the assessment of the decorative and protective properties coatings.

Table 19. Technological and operational properties of the finishing composition

Parameter name	Parameter value for developed composition
Compression strength, MPa	5.5
Adhesion strength, MPa	0.89
Frost resistance, cycles	35
Water-holding capacity, %	97.9
Water absorption, %	10.15
Water resistance	0.73
Shrinkage strain, %	0.024
Water vapour permeation μ, mg/m·t·Pa	0.049
Viability, h	1.5
Presence of cracks due to shrinkage	no cracks

Table 19 presents the technological and operational properties of the finishing composition and coatings based on it, which were based on the developed dry mix and the prototype composition

Figures 16 - 19 show photographs of samples of finishing compositions applied to concrete substrates before testing for frost resistance and after testing.

Figure 16. Appearance of the coating before testing.

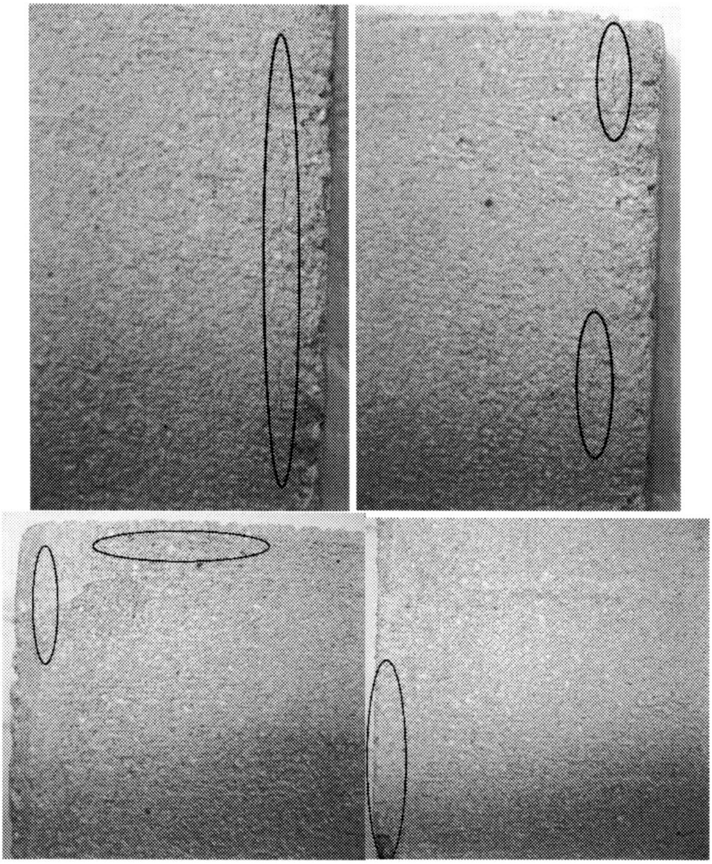

Figure 17. Appearance of the coating after 50 test cycles.

Figure 18. Composition W/B = 1.0, B: S = 1: 3, S-3, Neolith 7200, yellow pigment in an amount of 5% of the lime mass before testing.

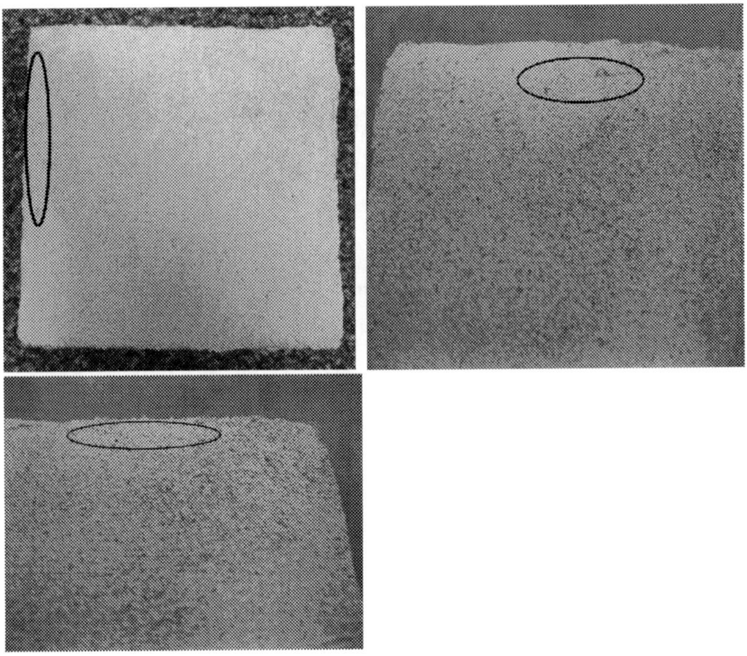

Figure 19. Composition W/B = 1.0, B: S = 1: 3, S-3, Neolith 7200, yellow pigment in an amount of 5% by weight of lime after testing.

Conclusion

For the first time, scientific data were obtained on the properties of the CSH filler and lime mixes with its use. Lime plaster composites with increased weather resistance have been developed, which is ensured by low values of open porosity and high values of strength. The resulting softening coefficient reaches 0.73, which is two times higher than the control unmodified composition. The following conclusions were drawn from the results of various tests:

The regularities of the filler synthesis were established depending on a density, and silicate modulus of liquid glass, the amount of the precipitant additive, the rate of its introduction, and the drying mode. The rational content of the precipitating additive was found to be 30–50% of the mass of liquid glass in the form of a 7.5–15% solution. The rational density of liquid glass was determined, which is 1130–1663 kg/m^3, depending on a silicate modulus of liquid glass. The filler output was 85–100%, depending on the synthesis conditions.

The phase composition of the filler, characterized by the presence of calcium silicate hydrates of the tobermorite group, a solid solution CSH (B) in the form of a weakly crystallized gel, a solid solution of C–S–H(II), hydrohalites, and calcites, was revealed. It was shown that the curves of the particle size distribution of the silicate-containing filler are of a polyfractional nature. The average particle diameter is 28–34 μm, depending on the filler synthesis mode. It was found that the filler based on calcium silicate hydrates has a high sorption capacity of 95% at 100% air humidity.

The chemical interaction of the CSH filler with lime was established. It was shown that the filler based on calcium silicate hydrates has a hydraulic activity of 178–289 mg/g, depending on the synthesis mode. It was found that an amount of free lime in the lime composites was reduced by more than two times in the presence of CSH additive.

It was revealed that lime compositions with CSH are characterized by a higher rate of strength development in comparison with the reference compositions. A mathematical model of the kinetics of hardening of a lime composite based on the binder filled with CSH was obtained. The regularities of the change in the strength of the lime composite depending on the fineness of the filler grinding, its content, and an amount of mixing water was established. It was revealed that the introduction of CSH to the lime composition leads to a decrease in porosity and an increase in the volume of closed pores of the composite.

The regularities of changes in the rheological and technological properties of lime finishing compositions were established depending on the conditions for the synthesis of the filler and its content, as well as the type of plasticizing additive. The introduction of the filler based on calcium hydro silicates into lime promotes a 13–21% increase in the effect of plasticizing additives. The use of superplasticizers S-3 or SP-3 gives the maximum plasticizing effect of the finishing lime composition. In this case, the water-reducing effect is 1.8.

The prospects for further research should be considered in the direction of the possibility of obtaining organic, and mineral additives for the modification of dry plaster mixes, expanding the range of local raw materials, and continuing research in the direction of studying the features of the structure formation processes of various finishing coatings.

References

[1] Dylewski, R., Adamczyk, J. (2014). The comparison of thermal insulation types of plaster with cement plaster. *J. Clean. Prod.*, 83, 256–262.

[2] De Azevedo, A.R.G., Alexandre, J., Zanelato, E.B., Marvila, M.T. (2017). Influence of incorporation of glass waste on the rheological properties of adhesive mortar. *Constr. Build. Mater.,* 148, 359–368.

[3] Groot, C., Van Hees, R., Wijffels, T. (2009). Selection of plasters and renders for salt laden masonry substrates. *Constr. Build. Mater.*, 23, 1743–1750.

[4] Murmu, M., Singh, S.P. (2015). Strength characteristics of lime-activated slag cement. *Adv. Cem. Res.*, 27, 268–277.

[5] Turanli, L., Saritas, A. (2011). Strengthening the structural behavior of adobe walls using plaster reinforcement mesh. *Constr. Build. Mater.*, 25, 1747–1752.

[6] Yardim, Y., Lalaj, O. (2016). Shear strengthening of unreinforced masonry wall with different fiber reinforced mortar jacketing. *Constr. Build. Mater.*, 102, 149–154.

[7] Zimich, V. (2016). Magnesium dry mixes for outer wall decoration. *Mater. Sci. Forum*, 870, 3–7.

[8] ˇCerný, R., Kunca, A.; Tydlitát, V., Drchalová, J., Rovnaníková, P. (2006). Effect of pozzolanic admixtures on mechanical, thermal and hygric properties of lime plasters. *Constr. Build. Mater*., 20, 849–857.

[9] Loganina, Valentina, Sergeeva, Kristina, Fediuk, Roman, Uvarov, Valery, Vatin, Nikolai, Vasilev, Yuriy, Amran, Mugahed, Szelag, Maciej. (2021). Increase the Performances of Lime Finishing Mixes due to Modification with Calcium Silicate Hydrates. *Crystals*, 11, no. 4: 399.

[10] Cultrone, G., Sebastián, E., Huertas, M.O. (2007). Durability of masonry systems: A laboratory study. *Constr. Build. Mater.*, 21, 40–51.

[11] Loganina, V., Sergeeva, K., Fediuk, R., Klyuev, S., Vatin, N. and Vasilev Y. (2021). Modified Lime Binders for Restoration Work. *Buildings*, 11, 98.

[12] D'Armada, P., Hirst, E. (2012). Nano-lime for consolidation of plaster and stone. *J. Arch. Conserv.*, 18, 63–80.

[13] Ibragimov, R., Fediuk, R. (2019). Improving the early strength of concrete: Effect of mechanochemical activation of the cementitious suspension and using of various superplasticizers. *Constr. Build. Mater.*, 226, 839–848.

[14] Pavlík, Z., Fort, J., Pavlíková, M., Pokorný, J.; Trník, A., ˇCerný, R. (2016). Modified lime-cement plasters with enhanced thermal and hygric storage capacity for moderation of interior climate. *Energy Build.*, 126, 113–127.

[15] Loganina, V.I., and Pyshkina, I.S. (2017). Optimization of synthesis mode for hydro silicate-based supplements used for lime dry construction mixtures. *Solid State Phenomena*, 265, pp. 9-13,

[16] Hauptmann, A., Yalcin, U. (2000). Lime plaster, cement and the first puzzolanic reaction. *Paléorient*, 26, 61–68.

[17] Loganina, V.I., Simonov, E.E., Jezierski, W., Małaszkiewicz, D. (2014). Application of activated diatomite for dry lime mixes. *Constr. Build. Mater.*, 65, 29–37.

[18] Fediuk, R.S., Lesovik, V.S., Liseitsev, Y.L., Timokhin, R.A., Bituyev, A.V., Zaiakhanov, M.Y., Mochalov, A.V. (2019). Composite binders for concretes with improved shock resistance. *Mag. Civ. Eng.* 2019.

[19] Popov, O., Moskalova, C. (2016). An analysis of the technological properties of facilitated plaster solutions made from dry building mixes. *Teh. Glas.*, 10, 102–105.

[20] Pahlavan, P., Manzi, S., Rodriguez-Estrada, M.T., Bignozzi, M.C. (2017). Valorization of spent cooking oils in hydrophobic waste-based lime mortars for restorative rendering applications. *Constr. Build. Mater.*, 146, 199–209.

[21] Hamard, E., Morel, J.-C., Salgado, F., Marcom, A., Meunier, N. (2013). A procedure to assess the suitability of plaster to protect vernacular earthen architecture. *J. Cult. Heritage*, 14, 109–115.

[22] Asaad, M.A., Ismail, M., Khalid, N.H.A., Huseien, G.F., Raja, P.B. (2018). Elaeis guineensis leaves extracts as eco-friendly corrosion inhibitor for mild steel in hydrochloric acid. *J. Teknol.*, 80.

[23] Loganina, V.I., Pyshkina, I.S. Martyashin G.V. (2018). Synthesis of supplement based on calcium hydro silicate for dry mixes. *Applied Mechanics and Materials*, 876) pp. 31-35.

[24] Vejmelková, E., Keppert, M., Rovnaníková, P., Keršner, Z., Cerny, R. (2012). Application of burnt clay shale as pozzolan addition to lime mortar. *Cem. Concr. Compos.*, 34, 486–492.

[25] Loganina, V.I., Makarova, L.V., Tarasov, R.V., Sergeeva, K.A. (2015). Increase water resistance decorative coatings on the basis of calcareous dry building mixtures. *Advanced Materials Research*, 1078, pp 171-173.

[26] Walker, R., Pavía, S. (2010). Physical properties and reactivity of pozzolans, and their influence on the properties of lime–pozzolan pastes. *Mater. Struct.*, 44, 1139–1150.

[27] Zegardło, B., Szelag, M., Ogrodnik, P. (2016). Ultra-high strength concrete made with recycled aggregate from sanitary ceramic wastes—The method of production and the interfacial transition zone. *Constr. Build. Mater.*, 122, 736–742.

[28] Pogorelov, S., Semenyak, G. (2016). Frost resistance of the steel fiber reinforced concrete containing active mineral additives. *Procedia Eng.* 150, 1491–1495.

[29] Pu, X. (1999) Investigation on pozzolanic effect of mineral additives in cement and concrete by specific strength index. *Cem. Concr. Res.*, 29, 951–955.

[30] Ezziane, K., Kadri, E.H., Hallal, A., Duval, R. (2009). Effect of mineral additives on the setting of blended cement by the maturity method. *Mater. Struct.*, 43, 393–401.

[31] Inozemtcev, A., Korolev, E., Qui, D.T. (2018). Study of mineral additives for cement materials for 3D-printing in construction. *IOP Conf. Ser. Mater. Sci. Eng.*, 365, 032009.

Chapter 4

Evaluation of the Effect of Simulated Tissue Fluid on Physical Properties of Four Calcium-Silicate Cements

I. Dimitrova*, **PhD**
and Y. Kouzmanova, PhD
Department of Conservative Dentistry,
Medical University, Sofia, Bulgaria

Abstract

Calcium-silicate cements (CSCs) are bioactive materials with a variety of applications in dentistry such as direct pulp capping, perforation repair, retrograde root-end filling, regenerative therapy, etc. Under clinical condition, these materials come into contact with tissue fluids and blood.

The aim of the present study was to evaluate and compare the effects of simulated tissue fluid (STF) on the Berkovich microhardness (HB) of four CSCs - white ProRoot MTA, gray MTA-Angelus, BioAggragate, and Biodentine. After setting, the samples were immersed in STF for different periods of time. Changes were assessed over a period of 24 hours and 30 days. Twelve measurements were made on each sample. A total of 80 samples were examined.

At the 24th hour Biodentine showed the greatest mean microhardness value, but a significant reduction in its microhardness was found from 0.74 GPa (±0.07) to 0.48 GPa (±0.07) after 30 days of exposure to STF. A significant increase of microhardness of MTA-Angelus and Bioaggregate was found at 30th day.

* Corresponding Author's Email: vanja_ves@abv.bg.

In: Properties and Uses of Calcium Silicate
Editor: Amanda G. Carlton
ISBN: 979-8-88697-128-6

Conclusion: The contact with STF for a period of 30 days led to significant changes of the microhardness of all tested materials, except for ProRoot MTA.

Keywords: calcium-silicate cements, microhardness, simulated tissue fluid

Introduction

The era of bioactive calcium-silicate cements in dental practice started in 1998 with the MTA Original-ProRoot® MTA (Dentsply, Tulsa Dental, Johnson City, TN, USA). This first cement was developed by the Torrebinead and Dr. White team (Bansal et al., 2020) and consisted of tricalcium silicate, tricalcium aluminate and dicalcium silicate and had a gray color of powder. In 2002, the white form ProRoot MTA was introduced. Although the two products have a similar composition, some differences between them exist such as a lower content of iron oxide in the formulation and significantly smaller particle size of the powder of the white form (Asgary et al., 2006). This bioactive material belongs to the first generation of the cements of MTA family.

A number of serious disadvantages of this originally introduced cement led to the development of many new brands in dental practice (Dutta et al., 2014). The main guidelines for changes in the new materials are as follows:

1. Modifications related to the shortening of the setting time by reducing the particle size of the powder composition, adding accelerators to the setting process, etc.
2. Modifications related to the reduction of the staining of tooth structure by replacing bismuth oxide with zirconium oxide, tantalum oxide, etc.
3. Modifications related to improving the handling properties by adding propylene glycol, water-soluble polymer, organic plasticizers, etc. (Dutta et al., 2014).

Basic Calcium-Silicate Cements Used in the Present Study: Brief Characteristics

In-depth knowledge of the characteristics of the materials used in dentistry, their composition, interactions and adverse chemical reactions are extremely important not only for their proper preparation, but also for a good outcome of treatment after their application.

MTA-Angelus (Angelus Soluçoes Odontológicas, Londrina, Brazil) has a similar composition to ProRoot MTA, but it does not contain calcium sulfate dehydrate (Dutta et al., 2014; Guven et al., 2014). This bioactive cement has been in dental practice since 2001 and belongs to the second generation. It is available in two forms: white and gray MTA. The white form does not contain iron (Camilleri, 2008). According to the manufacturers the setting time is within 15 minutes (Angelus Soluçoes Odontológicas, Londrina, Brazil) but according to Demirkaya et al. (2018) the setting time is within 10 minutes.

Bioaggregate (IBC Inc., Vancouver, Canada) is a bioactive nanomaterial with a nanoparticle size below 2 microns and with a white colored powder. Its composition is similar to that of white ProRoot MTA, but the differences are related to the higher content of phosphates in the composition and to the replacement of the radiopacifying agent - bismuth oxide with tantalum oxide (Park, J.W., et al. 2010; Saghiri MA, et al. 2013; L. Akinci et al. 2020). It was introduced in clinical practice in 2009 and refers to the third generation (Dutta et al. 2014). According to the manufacturers, the setting time is between 4-72 hours (DiaRoot Booklet, Features & Benefits, Verio Dental Co. Ltd., Vancouver, Canada).

Biodentine (Septodont, Saint-Maur-des-Fossés, France) is a third generation CSCs, introduced in 2010 and developed on the basis of active biosilicate technology as a dentin substitute (Biodentine ™ Active Biosilicate Technology. Scientific file, 2010; Debelian et al., 2016; Dutta et al., 2014). This cement has a chemical composition similar to white ProRoot MTA, but the radiopacifying agent bismuth oxide from their formula is replaced by zirconium oxide. The liquid represents an aqueous solution of calcium chloride and polycarboxylate (Nabeel et al., 2019).

Calcium-silicate cements are bioactive materials with various applications in dental practice such as: repair of perforations, direct pulp capping, treatment of resorptive defects, etc. (Camilleri et al., 2022) Under clinical conditions, these materials are in contact with various biological media such as: blood, tissue fluids, acidic environment, pus, various irrigants, etc. The available information in the literature concerning the influence of different media on the composition and some physicochemical properties of the biomaterials such as marginal adaptation, compressive strength, etc. is controversial and insufficient (Agrafioti et al., 2016; Susumu et al., 2020; do Prado et al., 2019; Prasanthi et al., 2019).

Analytical Framework

Literature Review

Influence of Environmental Conditions

There is no consensus in the literature regarding the influence of different factors and conditions of the contact environment on the different properties of the CSCs (Bolhari et al., 2015).

Influence of Environmental Conditions with Low Acidic pH

Inflammatory processes in the dental pulp and periodontium are associated with the presence of an acidic pH, as a result of the metabolic activity of microorganisms. The butyric acid is a major product of the metabolism of anaerobic microorganisms and was used in a study by Namazikhah and et al. (2008) in order to simulate the clinical situation of the inflammatory process in the periodontium. They found a significant reduction in the microhardness of MTA on contact with this acid with a pH of 4.4-7.4.

In a comparative study evaluating the influence of environment with different pH from 7.4, 6.4, 5.4, and 4.4 for a period of 7 days, Elnagby (2014) concluded that Biodentine is a much more suitable biomaterial compared to white MTA for application in conditions with acidic pH. The author registered changes in the microhardness, compressive strength and microstructure of all tested materials, as the established changes were more pronounced in wMTA. In another study Aggarwal et al. (2011) found that the combination of 17% EDTA and BioPure MTAD for 7 days significantly reduced the microhardness of white MTA.

According to Lee et al. (2004) environment with acidic pH leads to a significant deterioration of the hydration process of MTA. Yavari et al. (2013) found that an environment with increased acidity leads to increased solubility and porosity of the CSCs (2013). Similar results were found by Agrafioti et al. (2016). They did not observe changes in the sealing ability of both ProRoot and Biodentine, when these cements were in contact with citric acid at a pH of 5.4 for 24 hours. Additionally, the authors did not register changes for the 3-month period and concluded that these cements are particularly suitable for use in acidic conditions.

Contrary to that, Ballal et al. (2018) found significant changes, both in topography and in the microhardness of Biodentine, when it was in contact with 17% EDTA, 37% phosphoric acid, and 7% maleic acid for 1 min influence.

Effect of Blood Contamination

CSCs are used in endodontic practice for direct pulp capping, root perforation repair, root-end filling, etc. In all these applications the material comes into direct contact with a blood and tissue fluids. The parameters of the influence of blood and tissue fluids on the mechanical properties of these bioactive materials are extremely important, not only for their structure, but also for the proper outcome of healing processes in the long period.

In a study evaluating the bond strength of MTA-Angelus and Biodentine to the root dentin, Marquezan et al. (2018) compared treated with blood samples to untreated ones, during their hydration period at 24 hours, 7 and 28 days. They did not establish differences between treated and untreated Biodentine samples for the whole duration of the study. Moreover, they found changes in the bond strength of the treated with blood samples of MTA Angelus, but these changes were registered only in the first 24-hour period. In a comparative study on the resistance to dislodgement between blood treated groups and no blood treated Rahimi et al. (2013) found a significant reduction in the resistance of the bioactive material on the 7th day in the group, contaminated with blood. They concluded that the recovery time with restorative material is very important.

Ashofteh Yazdi et al. (2019) found that the contact of Biodentine with blood during its hydration process did not cause structural changes in the material and did not affect the amount of calcium hydroxide, while the blood exposure affected the hydration of ProRoot. Similar data were found by Dimond (2019). Through micro-CT evaluation, he noticed that the blood contamination of Biodentine, applied as root-end filling material did not affect its quality of sealing, but registered an increase in the volume of internal voids in the structure of the material. Using the same assessment method, Akinci, Simsek, and Aydinbelge (2020) found that the contact of Biodentine with fresh blood leads to a significant reduction of volume and density of the material.

Conversely, Salem Milani et al. (2013) observed that the contact with blood has a negative effect on the marginal adaptation of MTA. In addition, they found significant differences in the microstructure of MTA upon contact with blood and simulated tissue fluids (STF). Nekoofar et al. (2010a) found that when MTA becomes in contact with blood, its basic physical properties were significantly compromised. In contrast, Torabinejad et al. (1994) stated that the presence or absence of blood contamination had no significant effect on the amount of dye leakage. Mathew et al. (2016) reported that the sealing ability of Biodentine and MTA was not affected by the blood contact and the gap size values were even significantly lower than in the not blood treated groups.

Samiei et al. (2015) suggested that STF and stimulated plasma have very similar composition and effects to the natural tissue fluids and are particularly suitable for comparative studies on the properties of cements simulating clinical and natural environmental conditions.

Influence of Different Types of Irrigants

In clinical endodontic practice, sodium hypochlorite solutions with different concentrations, 2% chlorhexidine gluconate solution, and 17% ethylenediamine tetraacetic acid are widely used for disinfection and they are important parts of the irrigation treatment protocol. Palatyńska-Ulatowska et al. (2020) found that the use of 2% and 5.25% sodium hypochlorite with or without ultrasonic activation for different periods of time (12 minutes, 45 minutes and 24 hours) from the beginning of the setting process of Biodentine led to significant changes in its structure, but not in its chemical composition. They reported that 12 minutes were insufficient for complete hardening of Biodentine and to withstand the hypochlorite irrigation with and without ultrasonic activation. Moreover, the authors found that all samples were totally degraded. They emphasized the importance and key role of the hardening quality of cements before the use of irrigants.

Dasani et al. (Dasani et al. 2020) in a comparative study of the effect of 5% sodium hypochlorite solution and 17% EDTA for 10 minutes on the samples of Biodentine and MTA Angelus found that sodium hypochlorite solution and EDTA lead to a significant reduction of microhardness during their setting period. However, the authors report stronger changes in the samples treated with EDTA. Changes in microhardness are more pronounced at Biodentine. Thus, based on their results, the authors concluded that the used irrigants had an adverse effect not only on the microhardness of the tested cements, but also on their setting process.

Aggarwal et al. (Vivek Aggarwal, et al. 2011) found that the microhardness of white MTA decreased after exposure to 5% sodium hypochlorite and 2% chlorhexidine. According to Smith, J.B et al. (Smith, J.B et al. 2007) the effect of 1.3% sodium hypochlorite and the final irrigation with 17% EDTA for 5 min and or BioPure MTAD, applied for 1, 3 or 5 min on ProRoot MTA by three-dimensional profilometry analysis leads to significantly increased dissolution of the cement. The depth of the removed

MTA when applying EDTA and BioPure MTAD was within 1.5 and 8 μm. MTAD (a mixture of tetracycline, acid and a detergent) is an alternative irrigant designed to remove the smear layer, improving the sealing ability of Biodentine in a retrograde cavity (Naik, et al. 2015).

According to Singh et al. (2016) exposure to 3% sodium hypochlorite and 2% chlorhexidine gluconate for 30 minutes resulted in an increased resistance to dislodgement of Biodentine compared to MTA-Angelus under the same conditions. They concluded that chlorhexidine gluconate significantly reduced the resistance to dislodgement of MTA samples compared to those treated with hypochlorite.

Influence of Medications

According to Ozturk et al. (2019) the use of calcium hydroxide paste as an intracanal medicament reduced the gap size between ProRoot MTA and the root dentin wall, while the application for 1 or 4 weeks of triple and double antibiotic paste in regenerative therapy significantly impaired marginal adaptation of MTA and Biodentine. Similar results were established by Pereira et al. (2019). According to the authors, a mixture of calcium hydroxide paste and 2% chlorhexidine gel, used as an intracanal medication did not affect the bond strength between both Biodentine and white MTA-Angelus and root dentin wall, but the use of triple antibiotic paste significantly reduces the push-out bond strength. In contrast, Nagas et al. (2016) found that the resistance to dislodgement of Biodentine was less affected after treatment with Augmentin, triple antibiotic paste and Ledermix. Additionally, the authors reported that the use of calcium hydroxide paste significantly improved the adhesion of both Biodentine and ProRoot to root dentin walls.

Unlike them, Turk, Ozisik and Aydin (2015) reported a significant negative effect on the push-out bond strength after administration of calcium hydroxide as an intracanal medicament. In contrast, Topcuoglu et al. (2014) did not report a deterioration in the relationship between CSCs and dentin after the administration of calcium hydroxide and triple antibiotic paste. According to Jain et al. (2020) the marginal adaptation of Biodentine in the apical third of the root canal is significantly improved after exposure to a combination of calcium hydroxyl paste and propolis for 3 weeks compared to a mixture of triple antibiotic paste and Metapex for the same period and conditions.

Influence of the Experimental Immersion Media

Some material properties, such as solubility, surface micromorphology and elemental composition, have been shown to be dependent on the immersion media including water, simulated body fluid and Dulbecco's modified eagle medium (Benezra, Wismayer and Camilleri, 2017; Danesh et al., 2006). Using a new μCT-based method for calculating the solubility of four CSCs used as root-end filling materials after exposure to different immersion solutions - water, HBSS or HBSS+FCS, Camilleri et al. (2020) reported that it was dependent on the used solution, while the volume changes were not influenced by them.

Physical tests should replicate natural physiological conditions, i.e., 37°C. According to the accepted standards, when samples are tested, they should not be submerged in water but rather in a buffered saline solution, so that there are physiological concentrations of ions, and the testing conditions are more aligned to clinical conditions (Ha et al., 2017). Furthermore, the methodology used in the standard testing should be more relevant to clinical situations.

The impact of the foetal bovine serum (FBS) on physicochemical properties of MTA are assesses by several studies. Tingey, Bush and Levine (2008) recorded surface changes in hardened MTA upon contact with FBS. Kim et al. (2012) found that contact of both MTA-Angelus and ProRoot MTA with this serum interfered with their hydration. Kang et al. (2012) observed that the contact of MTA with FBS had an adverse effect on the microhardness of the cement. The choice of appropriate media in the experimental CSCs tests is important for the results obtained and the conclusions from them.

The review of the literature data clearly highlights the lack of definite unity in the opinions on the manner and type of impact and influence of different environmental conditions on the physico-chemical properties of CSC. And the role of the contact environment on the physico-chemical properties of CSC is extremely important not only for the bioactive materials themselves, but also for the proper course of the healing process after their application.

Information Withholding Hypothesis

Based on the literature review and on the differences in chemical compositions of the calcium silicate-based materials, the following *null hypotheses* were proposed: (1) The *first null hypothesis* stated that the microhardness values of all tested calcium-silicate cements would not be affected by exposure to simulated tissue fluid for a period of 30 days; (2) *The second null hypothesis* stated that there was a significant difference in the microhardness of the conventional MTA materials (ProRoot MTA and MTA-Angelus) compared to the newer CSCs (Bioaggregate and Biodentine).

Therefore, *the aim* of the present study was to evaluate and compare the effect of simulated tissue fluid on the Berkovich microhardness (HB) of four CSCs, such as white ProRoot MTA, gray MTA Angelus, BioAggragate, and Biodentine for a different period of time.

Method

Sample Construction

Four calcium silicate cements were investigated in this study, which were divided into 8 groups of 10 samples each.,The Berkovich microhardness (HB) of the materials was evaluated after immersion of the samples in simulated tissue fluid (STF) in two time intervals: on the 24th hour and on the 30th day.

Group I - White MTA ProRoot (Dentsply, Tulsa, Johnson City, USA) - 24th hour.
Group II - Gray MTA-Angelus (Angelus, Londrina, Brazil) - 24th hour.
Group III - BioAggregate (IBC Inc., Vancouver, Canada) - 24th hour.
Group IV - Biodentine (Septodont, Saint-Maur-des-Fossés, France) - 24th hour.
Group V - White MTA ProRoot (Dentsply, Tulsa, Johnson City, USA) - 30th day.
Group VI - Gray MTA-Angelus (Angelus, Londrina, Brazil) - 30th day.
Group VII - BioAggregate (IBC Inc., Vancouver, Canada) - 30th day.
Group VIII - Biodentine (Septodont, Saint-Maur-des-Fossés, France) - 30th day.

The materials were condensed in 80 polyethylene cylindrical molds with an inner diameter of 5 mm and a height of 3 mm. All calcium-silicate cements were prepared according to the manufacturer's instructions. ProRoot MTA, MTA-Angelus and BioAggregate were mixed by hand with distilled water on the smooth side of a glass slab with the powder/liquid ratio 3:1. Biodentine was prepared in an amalgamator. Five drops of its liquid were added to predosed capsule that was placed on a capsule mixer GC Silver Mix (Dubai Medical Equipments, UAE) at a speed of 4000 rpm for 30 seconds.

The polyethylene molds were placed on a glass slab, which served as their basis. The cements are applied in increments and condensed to a slight overfill.

Excess material on their upper side is removed and the surface was smoothed with a glass slap. A cotton pellet moisturized with saline solution was placed on this surface for 4 hours to provide the necessary additional moisture for the normal course of the hydration and setting process.

Thereafter, in order to mimic clinical conditions, the samples were arranged in Oasis floral foam (Smithers-Oasis comp.) and subsequently immersed in small quantity of simulated tissue fluid (Merck, Germany) so that their upper surface remained above the liquid level. The STF contained NaCl, KCl, Na_2HPO_4 and KH_2PO_4 and its ionic composition and pH (7.4) were close to those of the blood serum. The specimen preparation methods followed those of the previous study by Kim et al. (2012).

Microhardness Assessment by Berkovich Method

At the 24th hour, the samples from the first four groups were removed from the liquid, carefully rinsed with 10 ml of distilled water and dried on paper. They were afterwards fixed on metal cylinders and the surface which was in contact with STF was leveled and polished on a Leika microtome (Singapore) with a static glass knife. The microhardness of the materials was measured by the Berkovich method using a UNMT nano- and microindentation apparatus (Bruker, USA) combined with an atomic force microscope. The tip of the indenter was a diamond three-sided pyramid.

The indentation process involved several successive actions: penetration of the indenter into the material until a maximum load of 500 mN is reached, retention of this load on the sample for 10 seconds, unloading up to 10% of the load, retention for 15 seconds and finally, releasing the indenter until it exits the surface of the material. An impression with three edges remained on its surface. The testing conditions are represented in Figure 1.

For each sample, 12 measurements were made at a distance of 0.08 μm. The entire process was computerized. The average sample value was evaluated and measured by the relevant software program (UNMT Viewer) (Figure 2). These values were calculated automatically based on the records of the depth of penetration and the area of the trace in units of measurement of microhardness by Berkovich (HB). Likewise, the mean values of the material microhardness for each group were then calculated in GPa.

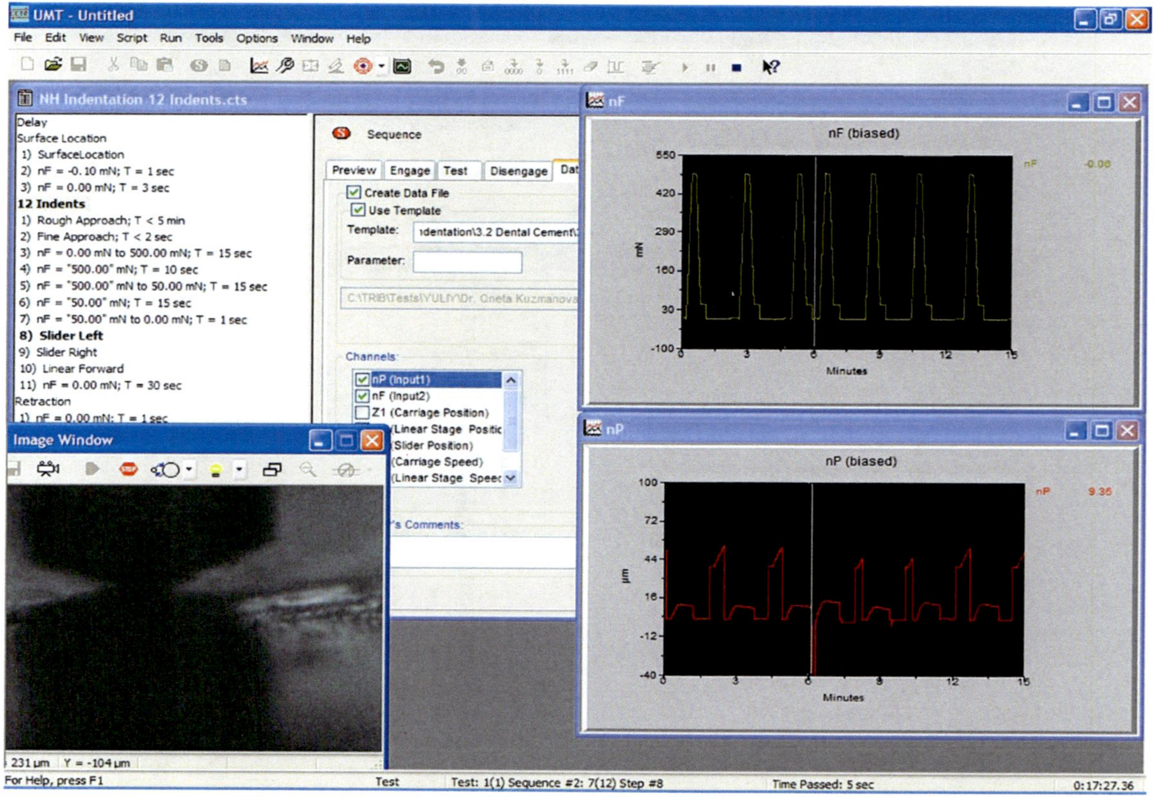

Figure 1. A screenshot representing the testing conditions during the microindentation process.

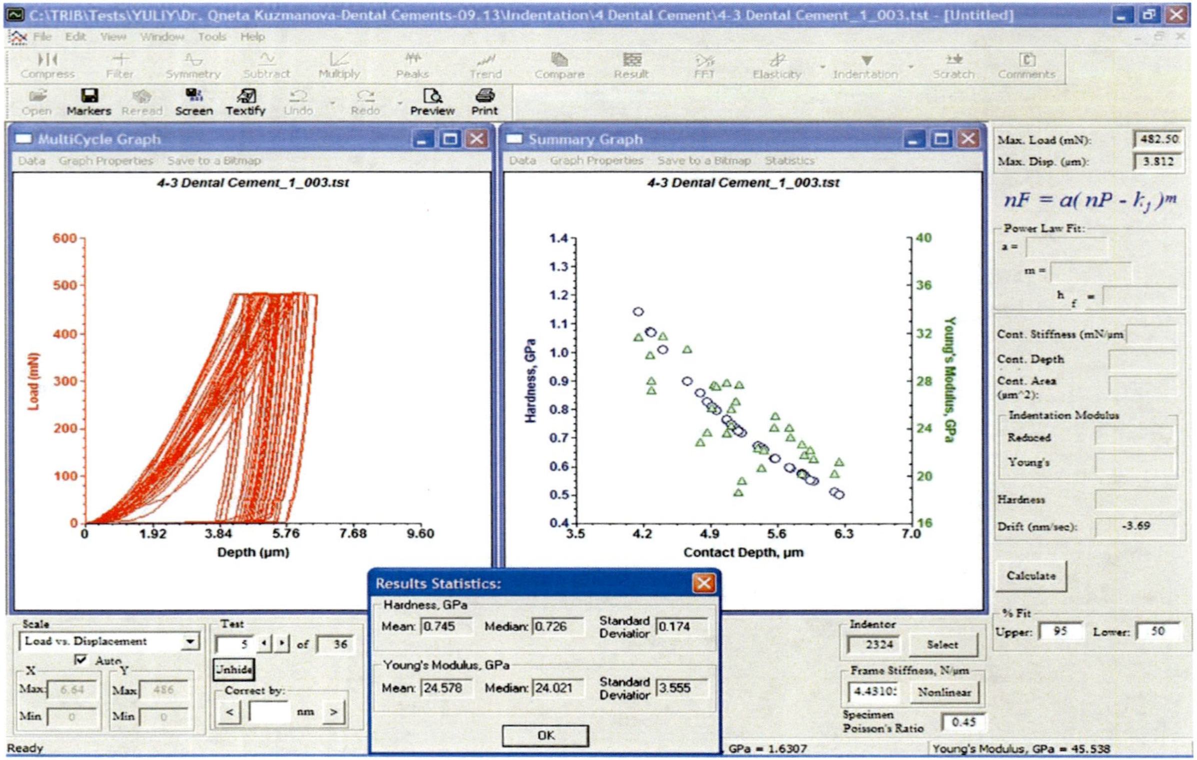

Figure 2. A representative screenshot for the computerized measurement and calculation of the mean microhardness value for each sample by the Berkovich method.

The samples from the remaining 4 groups (groups V-VIII) are stored under the same conditions for a period of 30 days, and the STF are refreshed daily. After removing them from the media, the tests described above were performed on them as well.

Statistical Analysis

All data are entered and processed with the statistical package IBM SPSS Statistics 25.0 (SPSS, Inc. Chikago, IL). Descriptive statistics (mean, standard error, standard deviation, confidence interval) is used to evaluate the microhardness of the materials. The results are also submitted to the following tests: (1) nonparametric tests of Shapiro-Wilk and Kolmogorov-Smirnov to check the normality of the distribution of the data; (2) Mann-Whithey U test and Wilcoxon test for comparison between two independent samples; (3) graphical analysis to visualize the results. α-level was set at 0.05 and P-value less than 0.05 ($p<0.05$) was considered statistically significant.

Results and Discussion

Descriptive Statistics

In this study was evaluated the effect of STF on the microhardness values of 4 calcium-silicate cements for two time periods: 24 hours and 30 days. The results are presented in Table 1 and Figure 3. At the 24th hour, Biodentine showed the highest microhardness (0.74 ± 0.07 GPa), followed by white ProRoot MTA (0.58 ± 0.14 GPa) and gray MTA-Angelus (0.18 ± 0.02 GPa), with statistically significant difference between all tested materials ($p<0.05$).

On the 30th day, white ProRoot MTA reached the highest microhardness value (0.66 ± 0.30 GPa) but the difference between it and both gray MTA-Angelus (0.54 ± 0.06 GPa) and Biodentine was insignificant. Generally, a slight increase of the average microhardness values was registered over time with almost all tested materials, except for Biodentine. The value of this material was reduced to 0.48 ± 0.07 GPa, with a statistically significant difference from that observed at the 24th hour. gMTA-Angelus showed the highest rate of increase of the mean microhardness value, with statistically significant difference, as seen in Figure 3. In addition, BioAggregate exhibited the lowest microhardness at the end of both time intervals (0.06 ± 0.02 GPa and 0.16 ± 0.04 GPa, respectively), with a statistically significant difference with the other three materials.

Table 1. Mean values of microhardness (in GPa) of CSCs for two periods of time

Time	Materials	N	Mean	SEM	95% CI	SD	Min	Max
24th hour	wPR	10	0.58^{a}	0.04	0.48-0.68	0.14	0.38	0.75
	gAG	10	0.18^{b}	0.02	0.13-0.23	0.06	0.07	0.26
	BA	10	0.06^{c}	0.01	0.05-0.07	0.02	0.04	0.09
	BD	10	0.74^{d}	0.02	0.70-0.79	0.07	0.63	0.82
30th day	wPR	10	0.66^{ad}	0.09	0.47-0.89	0.30	0.14	0.99
	gAG	10	0.54^{a}	0.06	0.40-0.68	0.96	0.33	0.91
	BA	10	0.16^{e}	0.01	0.13-0.19	0.04	0.12	0.26
	BD	10	0.48^{a}	0.02	0.43-0.53	0.07	0.32	0.60

Significant variables at levels of significance ($p<0.05$) are respectively expressed by different letters. Legend: wPR – white ProRoot MTA, gAG – grey MTA-Angelus, BA – Bioaggregate, BD – Biodentine; SEM - Standard Error of the Mean; CI – Confident Interval; SD - Standard Deviation.

As mentioned above, in the present study Biodentine showed the greatest mean microhardness value at the 24th hour, but a statistically significant reduction in its microhardness was found from 0.74 GPa (±0.07) to 0.48 GPa (±.07) on the 30th day of contact with a simulated tissue fluid ($p<0,05$). Contrary to that, gray MTA-Angelus and Bioaggregate showed significant increase of their average value. Such a difference was not established in ProRoot MTA group. Therefore, the first null hypothesis tested was partially rejected and accepted only in regard to ProRoot MTA. Moreover, significant differences were found between the average values of this physical property in the four main groups of materials. Hence, the second null hypothesis was accepted. Additionally, a visible change in color towards graying was observed in white ProRoot MTA group after a period of 30 days, while the other three CSCs did not show a discoloration.

Blood plasma in the human body represents 55% of the total blood volume and is composed of 90% water, sodium salts, potassium, phosphates and sulfates as lipids and proteins (Kokubo et al., 1990). The simulated tissue fluid used in this study, as well as by other authors (Reyes-Carmona, Felippe, and Felippe, 2010; Saghiri et al., 2010), is a phosphate buffered saline with pH 7.4 and contains NaCl, KCl, Na_2HPO_4, and KH_2PO_4. This kind of STF may mimic the inorganic ionic composition of blood plasma but without lipids and proteins in its formulation (Innes et al., 2021; Marques et al., 2011).

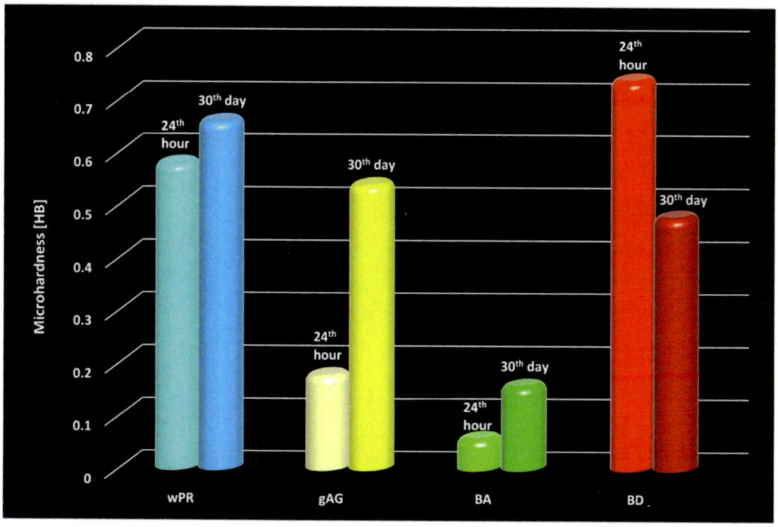

Figure 3. Comparison between the Berkovich microhardness values (in GPa) of four CSCs at two periods of time.

According to Camilleri, Formosa, and Damidot (2013), the contact of MTA with tissue fluids such as phosphates impairs the hydration process and prolongs the time of its hardening. They detected the presence of microcracks, as a result of volumetric expansion of cement and the formation of insoluble calcium phosphate.

The tests for microhardness of the endodontic materials fall into two types of indentations: (1) microindentation as determined by the Knoop, Vickers, and Berkovich indenters; and (2) nanoindentation as determined by the Berkovich indenter. These methods are based on the resistance to plastic deformation of the surface of the test material after indenter penetration (Geogi et al., 2022; Chandler, 1999). To the best of the authors' knowledge, only one study performed nanoindentation Berkovich tests for assessment of the microhardness of ProRoot MTA (Žižka et al., 2018). In the present study, the microhardness of different CSCs after exposure to STF using nanoindentation was investigated for the first time.

Bolhari et al. (2015) reported lower microhardness values with Bioaggregate compared to ProRoot MTA. Similar results were found by Grech, Mallia, and Camilleri (2013), and according to them the microhardness of Bioaggregate was significantly lower than that of Biodentine. These findings are consistent with ous. Herein we report the lowest values of Bioaggregate microhardness in contact with STF for both time periods.

In the present study, the microhardness values of all tested CSCs, except for that of Biodentine, increased slightly during their 30-day contact with STF. In contrast, Antonijević et al. (2016) observed high value of microhardness of Biodentine in STF contact. Niu et al. (2014) suggested that the possible explanation of this fact could be attributed to the strong crystal growth, found only in Biodentine (2014).

In almost all areas of application, CSCs are in contact with blood and/or tissue fluids, and the opinions regarding their effects on the physicochemical properties of cements are contradictory in the literature. Nekoofar et al. (2010a) reported that mixing MTA with blood has a negative effect on its microhardness. The compressive strength of white ProRoot MTA was higher than those of gray form in the presence of blood and tissue fluids, and some changes in its surface microstructure were found (Nekoofar et al. 2010; Nekoofar, Stone, and Dummer, 201b; Nekoofar et al., 2011).

In the conditions of the present study only the white ProRoot MTA microhardness values were relatively high at both time periods: at 24th hour and after 30 days. Unlike other cements, Biodentine samples exhibited a significant reduction in microhardness after exposure to STF for 30 days compared to the microhardness of the samples assessed at 24th hour. According to Meraji (2014), the effect of synthetic phosphate solutions on CSCs improved their physical properties, while the acidic pH of the medium had a negative effect on their microhardness and compressive strength. Similar data are found by other authors (Kayahan et al., 2009; Namazikhah et al., 2008).

The highest initial microhardness values recorded by us at the 24th hour with the Biodentine samples may be due to increased moisture conditions, which favor the setting process. However, the unexpected decrease of its microhardness after 30 days may be a result of the prolonged contact with STF and possible chemical interaction with it.

Another explanation of the reduction of microhardness values for Biodentine on the 30th day when compared to the other tested CSCs, could be attributed to the adverse effect of the STF on the hardening process and disruption of its hydration. There is no consensus in the literature about the exact initial and final setting time of Biodentine. The duration of the setting time of calcium-silicate biomaterials is one of the most important factors that influence their properties. Thus, for example, according to Atmeh and Watson (2022), the setting time of Biodentine varies from 6.5 to 86 minutes. These data, obtained through controlled laboratory studies, differ from clinical ones.

According to the manufacturers and other authors, the setting time of this cement is within 10-12 minutes (Biodentine ™ Active Biosilicate Technology, 2010; Jagadish and Krishna, 2019). Ha et al. (2017) reported that according to the ISO 9917.1 standard, the settting time of Biodentine is within 45-85.7 minutes. According to other authors (Alhodiry, Lyons, and, Chadwick, 2014; Moosani et al., 2007; Rajasekharan et al., 2018), under the influence of blood its setting time can vary from 16 minutes to 6 hour.

Incomplete setting of Biodentine and subsequent contact with STF may lead to an increase in its solubility and porosity. This may be due to the fact that blood leads to an increase in the porosity of Biodentine. Similar findings were established by Sheykhrezae et al. (2018). Our previous study regarding the solubility of five CSCs over a period of 28 days showed that Biodentine had the highest established solubility - 18.72% (Kouzmanova and Dimitrova, 2020).

According to Niu et al. (2014), CSCs as bioactive materials interact with the contact environment, which may affect their composition, structure, and properties. A number of authors reported a negative effect of blood on the properties of CSCs (Song et al., 2016; Üstün et al., 2015; Yavus and Eyüboğlu, 2021). However, there is evidence in the literature about the lack of effect of blood on MTA properties (Hernandez et al., 2005; Kogan et al., 2006).

And finally, Cammilleri et al. (2022) reported no specification for standard testing of the solubility of CSCs. The status quo is the same for their hardness testing. In our opinion, the most suitable method for measuring the microhardness of nanomaterials such as Bioaggregate is the nanoindentation Berkovich test.

Conclusion

This research examined the effect of simulated tissue fluid on the microhardness (HB) of four calcium silicate-based cements for two periods of time. Under the conditions of this laboratory evaluation and within its limitations, it can be concluded that the contact with STF for a period of 30 days leads to a significant changes in microhardness of almost all tested materials. Gray MTA-Angelus and Bioaggregate showed an increase of their microhardness, unlike Biodentine, which exhibited a significant reduction in its average value. ProRoot MTA exhibited the highest mean value of this property on the 30th day, hence, it is still the material of choice for many endodontic applications.

References

Aggarwal, V., Jain, A. and Kabi, D. (2011). In vitro evaluation of effect of various endodontic solutions on selected physical properties of white mineral trioxide aggregate. *Australian Endodontic Journal*, 61-64.

Agrafioti, A., Tzimpoulas, N., Chatzitheodoridis, E., Kontakiotis, E. G. (2016). Comparative evaluation of sealing ability and microstructure of MTA and Biodentine after exposure to different environments. *Clinical Oral Investigations*, 1535-1540.

Akinci, L., Simsek, N., Aydinbelge, H. A. (2020). Physical properties of MTA, BioAggregate and Biodentine in simulated conditions: A micro-CT analysis. *Dental Materials Journal*, 601-607.

Alhodiry, W., Lyons, M. F. and Chadwick, R. G. (2014). Effect of saliva and blood contamination on the bi-axial flexural strength and setting time of two calcium-silicate based cements: Portland cement and biodentine. *The European Journal of Prosthodontics and Restorative Dentistry*, 20-23.

Al-Zubaidi, A. K. and Al-Azzawi, A. K. J. (2014). The effect of various endodontic irrigants on the sealing ability of Biodentine and other root perforation repair materials (In vitro study). *Journal of Baghdad College of Dentistry*, 1-8.

Angelus. *MTA Angelus*. Available from: https://angelus.ind.br/produt o/mta-angelus.

Antonijević, D., Milovanović, P., Riedel, C., Hahn, M., Amling, M., Busse, B. and Djurić, M. (2016). Application of reference point indentation for micro-mechanical surface characterization of calcium silicate based dental materials. *Biomedical Microdevices*, 25.

Asgary, S., Parirokh, M., Eghbal, M. J., Stowe, S. and Brink, F. (2006). A qualitative X-ray analysis of white and grey mineral trioxide aggregate using compositional imaging. *Journal of Materials Science. Materials in Medicine*, 187-191.

Ashofteh Yazdi, K., Ghabraei, S., Bolhari, B., Kafili, M., Meraji, N., Nekoofar, M. H. and Dummer, P. M. H. (2019). Microstructure and chemical analysis of four calcium silicate-based cements in different environmental conditions. *Clinical Oral Investigations*, 43-52.

Atmeh, A. R. and Watson, T. F. (2022). Biodentine™. Physico-chemical properties: From Interactions with dental tissues to Ageing. In: *Biodentine™. Properties and Clinical Applications*. About, I. Springer Cham, 11-30.

Ballal, V., Marques, J. N., Campos, C. N., Lima, C. O., Simão, R. A. and Prado, M. (2018). Effects of chelating agent and acids on Biodentine. *Australian Dental Journal*, 170-176.

Bansal, K., Jain, A., Aggarwal, N. and Jain, A. (2020). Biodentine VS MTA: A comparitive analysis. *International Journal of Oral Health Dentistry*, 201-208.

Bayraktar, K., Basturk, F. B., Turkaydin, D. and Gunday, M. (2021). Long-term effect of acidic pH on the surface microhardness of ProRoot mineral trioxide aggregate, Biodentine, and total fill root repair material putty. *Dental Research Journal (Isfahan)*, 2.

Benezra, M. K., Wismayer, P. S. and Camilleri, J. (2017). Influence of environment on testing of hydraulic sealers. *Scienticic Reports*, 17927.

Bolhari, B., Ashofteh Yazdi, K., Sharifi, F. and Pirmoazen, S. (2015). Comparative scanning electron microscopic study of the marginal adaptation of four root-end filling materials in presence and absence of blood. *Journal of Dentistry (Tehran)*, 226-234.

Bolhari, B., Nekoofar, M. H., Sharifian, M., Ghabrai, S., Meraji, N. and Dummer, P. M. (2014). Acid and microhardness of mineral trioxide aggregate and mineral trioxide aggregate-like materials. *Journal of Endodontics*, 432-435.

Camilleri, J. (2008). The chemical composition of mineral trioxide aggregate. *Journal of Conservative Dentistry*, 141-143.

Camilleri, J., Atmeh, A., Li, X. and Meschi, N. (2022). Present status and future directions: Hydraulic materials for endodontic use. *International Endodontic Journal*, 710-777.

Camilleri, J., Formosa, L. and Damidot, D. (2013). The setting characteristics of MTA Plus in different environmental conditions. *International Endodontic Journal*, 831-840.

Camilleri, J., Wang, C., Kandhari, S., Heran, J. and Shelton R. M. (2022). Methods for testing solubility of hydraulic calcium silicate cements for root-end filling. *Scientific Reports*, 7100.

Chandler, H. (1999). *Hardness Testing*. 2nd Ed., Materials Park, OH: ASM International, 1-13.

Geogi, C. C., Dubey, S., Singh, P. and Radjkumar, B. (2022). MTA Vs Biodentine. In: *Bioceramics in Endodontics*, Bookrivers, Lucknow, 100-104.

Grech, L., Mallia, B. and Camilleri, J. (2013). Investigation of the physical properties of tricalcium silicate cement-based root-end filling materials. *Dental Materials*, e20-28.

Danesh, G., Dammaschke, T., Gerth, H. U., Zandbiglari, T. and Schäfer, E. (2006). A comparative study of selected properties of ProRoot mineral trioxide aggregate and two Portland cements. *International Endodontic Journal*, 213-219.

Dasani, S., Hora, B. S., Ajmera, R. S., Gupta, B., Dondulkar, Y. and Kosta, S. (2021). The effect of 5% NaOCl and 17% EDTA, at 24hrs and 8days, on the microhardness of MTA, Biodentine and Pozzolan cement: An in-vitro study. *University Journal of Dental Sciences*, 16-21.

Debelian, G. and Trope, M. (2016). The use of premixed bioceramic materials in endodontics. *Giornale Italiano di Endodonzia*, 70-80.

Demirkaya, K., Ersahan, S., Suyun, G. and Aktürk, S. (2018). Evaluation of micro surface structure and chemical composition of two different calcium silicate–containing filling materials. *Turkish Endodontic Journal*, 10-14.

DENTSPLY Tulsa Dental Specialties. *ProRoot MTA (Mineral Trioxide Aggregate) Root Canal Repair Material*. Available from: https://www.dentsplyestore.com.au/secure/downloadfile.asp?fileid=1683397.

DiaDent Group International. DiaRoot® BioAggregate. *DiaRoot Booklet, Features & Benefits*. Available from: www.diadent.com.

Dimond, C. A. (2019). The effect of blood contamination on the retrograde sealing ability of Biodentine®: A micro-CT analysis [Thesis, University of Minnesota.]

do Prado, M., de Lima, C. O., das Neves Marques, L., Ballal, V., Renata Simão, A. and Campos, C. N. (2019). Effect of different irrigation regimens on the bond strength of Biodentine used for furcal perforation repair. *ENDO – Endodontic Practice Today*, 49-59.

Dutta, A. and Saunders, W. Calcium silicate materials in endodontics. *Dental Update*, 708-722.

Elnaghy, A. M. (2014). Influence of acidic environment on properties of biodentine and white mineral trioxide aggregate: a comparative study. *Journal of Endodontics*, 953-957.

Fridland, M. and Rosado, R. (2003). Mineral trioxide aggregate (MTA) solubility and porosity with different water-to-powder ratios. *Journal of Endodontics*, 814-817.

Guven, Y., Tuna, E. B., Dincol, M. E. and Aktoren, O. (2014). X-ray diffraction analysis of MTA-Plus, MTA-Angelus and DiaRoot BioAggregate. *European Journal of Dentistry*, 211-215.

Ha, W. N., Nicholson, T., Kahler, B. and Walsh, L. J. (2017). Mineral trioxide aggregate- A review of properties and testing methodologies. *Materials (Basel)*, 1261.

Han, L., Okiji, T. and Okawa, S. (2010). Morphological and chemical analysis of different precipitates on mineral trioxide aggregate immersed in different fluids. *Dental Materials Journal*, 512-517.

Innes, E., Yiu, H. H. P., McLean, P., Brown, W. and Boyles, M. (2021). Simulated biological fluids – a systematic review of their biological relevance and use in relation to inhalation toxicology of particles and fibres. *Critical Reviews in Toxicology*, 217-248.

Hernandez, E. P., Botero, T. M., Mantellini, M. G., McDonald, N. J. and Nör, J. E. (2005). Effect of ProRoot MTA mixed with chlorhexidine on apoptosis and cell cycle of fibroblasts and macrophages in vitro. *International Endodontic Journal*, 137-143.

Islam, I., Chng, H. K. and Yap, A. U. (2006). Comparison of the physical and mechanical properties of MTA and portland cement. *Journal of Endodontics*, 193-197.

Jagadish, L. and Krishna, A. (2019). Biodentine: A contemporary dentin deputy and boon to the field of dentistry. *IOSR Journal of Dental and Medical Sciences*, 68-72.

Jain, R. N., Abraham, S., Karad, R. R., Najan, H. B., Vaswani, S. D. and Torris, A. (2020). Micro-computed tomographic analysis of the marginal adaptation of a calcium silicate-based cement to radicular dentin after removal of three different intracanal medicaments – An in vitro study. *Journal of Conservative Dentistry*, 598-603.

Kang, J. S., Rhim, E. M., Huh, S. Y., Ahn, S. J., Kim, D. S., Kim, S. Y. and Park, S. H. (2012). The effects of humidity and serum on the surface microhardness and morphology of five retrograde filling materials. *Scanning*, 207-214.

Kayahan, M. B., Nekoofar, M. H., Kazandağ, M., Canpolat, C., Malkondu, O., Kaptan, F. and Dummer P. M. (2009). Effect of acid-etching procedure on selected physical properties of mineral trioxide aggregate. *International Endodontic Journal*, 1004-1014.

Kim, Y., Kim, S., Shin, Y. S., Jung, I. Y. and Lee, S. J. (2012). Failure of setting of mineral trioxide aggregate in the presence of fetal bovine serum and its prevention. *Journal of Endodontics*, 536-540.

Kogan, P., He, J., Glickman, G. N. and Watanabe, I. (2006). The effects of various additives on setting properties of MTA. *Journal of Endodontics*, 569-572.

Kokubo, T., Kushitani, H., Sakka, S., Kitsugi, T. and Yamamuro, T. (1990). Solutions able to reproduce in vivo surface-structure changes in bioactive glass-ceramic A-W. *Journal of Biomedical Materials Research*, 721-734.

Kouzmanova, Y., Dimitrova, I. (2020). Solubility of calcium silicate based cements – A comparative study. *Acta Medica Bulgarica*, 27-29.

Lee, Y. L., Lee, B. S., Lin, F. H., Yun Lin, A., Lan, W. H. and Lin, C. P. (2004). Effects of physiological environments on the hydration behavior of mineral trioxide aggregate. *Biomaterials*, 787-93.

Marques, M. R. C., Loebenberg, R. and Almukainzi, M. (2011). Simulated biological fluids with possible application in dissolution testing. *Dissolution Technologies*, 15-28.

Marquezan, F. K., Kopper, P. M. P., Dullius, A. I. D. S., Ardenghi, D. M. and Grazziotin-Soares, R. (2018). Effect of blood contamination on the push-out bond strength of calcium silicate cements. *Brazilian Dental Journal*, 189-194.

Mathew, L. A., Kini, S., Acharya, S. R., Kamath, S., Menezes, N. D. and Rao, S. (2016). A comparative evaluation of the microleakage of blood-contaminated mineral trioxide aggregate and Biodentine as root-end filling materials: An in vitro study. *Journal of Interdisciplinary Dentistry*, 19-24.

Meraji, N. (2014). Mixing MTA with phosphate buffered solution can improve its physical properties and increase favorable outcome. *Dental Hypotheses*, 14-15.

Moosani, G. K. R., Manduri, C. S., Sampathi, N. R., Katta, D. S. K., Medepalli, R., Sakaray, M., Marella, K. and Magam, P. K. (2007). Evaluation of setting time of mineral trioxide aggregate and Biodentine in the presence of human blood and minimal essential media – An in vitro study. *Journal of Evidence Based Medicine and Healthcare*, 5849-5852.

Nabeel, M., Tawfik, H. M., Abu-Seida, A. M. A. and Elgendy, A. A. (2019). Sealing ability of Biodentine versus ProRoot mineral trioxide aggregate as root-end filling materials. *Saudi Dental Journal*, 16-22.

Nagas, E., Cehreli, Z. C., Uyanik, M. O., Vallittu, P. K. and Lassila, L. V. (2016). Effect of several intracanal medicaments on the push-out bond strength of ProRoot MTA and Biodentine. *International Endodontic Journal*, 184-188.

Naik, M. M., de Ataide Ide, N., Fernandes, M. and Lambor, R. (2015). Assessment of apical seal obtained after irrigation of root end cavity with MTAD followed by subsequent retrofilling with MTA and Biodentine: An in vitro study. *Journal of Conservative Dentistry*, 132-135.

Namazikhah, M. S., Nekoofar, M. H., Sheykhrezae, M. S., Salariyeh, S., Hayes, S. J., Bryant, S. T., Mohammadi, M. M. and Dummer, P. M. (2008). The effect of pH on surface hardness and microstructure of mineral trioxide aggregate. *International Endodontic Journal*, 108-116.

Nekoofar, M. H., Davies, T. E., Stone, D., Basturk, F. B. and Dummer, P. M. (2011). Microstructure and chemical analysis of blood-contaminated mineral trioxide aggregate. *International Endodontic Journal*, 1011-1018.

Nekoofar, M. H., Oloomi, K., Sheykhrezae, M. S., Tabor, R., Stone, D. F. and Dummer, P. M. (2010a). An evaluation of the effect of blood and human serum on the surface microhardness and surface microstructure of mineral trioxide aggregate. *International Endodontic Journal*, 849-858.

Nekoofar, M. H., Stone, D. F. and Dummer, P. M. (2010b). The effect of blood contamination on the compressive strength and surface microstructure of mineral trioxide aggregate. *International Endodontic Journal*, 782-791.

Niu, L.-N., Jiao, K., Wang, T.-D., Zhang, W., Camilleri, J., Bergeron, B. E., Feng, H.-L., Mao, J., Chen, J.-H., Pashley, D. H. and Tay, F. R. (2014). A review of the bioactivity of hydraulic calcium silicate cements. *Journal of Dentistry*, 517-533.

Ozturk, T. Y., Guneser, M. B., Taschieri, S., Maddalone, M., Dincer, A. N., Venino, P. M. and Del Fabbro, M. (2019). Do the intracanal medicaments affect the marginal adaptation of calcium silicate-based materials to dentin? *Journal of Dental Sciences*, 157-162.

Palatyńska-Ulatowska, A., BuŁa, K. and Klimek, L. (2020). Influence of sodium hypochlorite and ultrasounds on surface features and chemical composition of Biodentine tricalcium silicate-based material. *Dental Materials Journal*, 587-592.

Park, J. W., Hong, S. H., Kim, J. H., Lee, S. J. and Shin, S. J. (2010). X-Ray diffraction analysis of white ProRoot MTA and Diadent BioAggregate. *Oral Surgery, Oral Medicine, Oral Pathology, Oral Radiology, Endodontics*, 155-158.

Pereira, A. C., Pallone, M. V., Marciano, M. A., Cortellazzi, K. L., Frozoni, M., Gomes, B. P. F. A., de Almeida, J. F. A. and Soares, A. J. (2019). Effect of intracanal medications on the interfacial properties of reparative cements. *Restorative Dentistry & Endodontics*, e21.

Prasanthi, P., Garlapati, R., Nagesh, B., Sujana, V., Kiran Naik, K. M. and Yamini, B. (2019). Effect of 17% ethylenediaminetetraacetic acid and 0.2% chitosan on pushout bond strength of biodentine and ProRoot mineral trioxide aggregate: An in vitro study. *Journal of Conservative Dentistry*, 387-390.

Rahimi, S., Ghasemi, N., Shahi, S., Lotfi, M., Froughreyhani, M., Milani, A. S. and Bahari, M. (2013). Effect of blood contamination on the retention characteristics of two endodontic biomaterials in simulated furcation perforations. *Journal of Endodontics*, 697-700.

Rajasekharan, S., Martens, L. C., Cauwels, R. G. E. C. and Anthonappa, R. P. (2018). Biodentine™ material characteristics and clinical applications: a 3 year literature review and update. *European Archives of Paediatric Dentistry*, 1-22.

Reyes-Carmona, J. F., Felippe, M. S. and Felippe, W. T. (2010). The biomineralization ability of mineral trioxide aggregate and Portland cement on dentin enhances the push-out strength. *Journal of Endodontics*, 286-291.

Saghiri, M. A., Garcia-Godoy, F., Asatourian, A., Lotfi, M., Banava, S. and Khezri-Boukani, K. (2013). Effect of pH on compressive strength of some modification of mineral trioxide aggregate. Medicina Oral, Patología Oral y Cirurgía Bucal, e714-20.

Saghiri, M. A., Shokouhinejad, N., Lotfi, M., Aminsobhani, M. and Saghiri, A. M. (2010). Push-out bond strength of mineral trioxide aggregate in the presence of alkaline pH. *Journal of Endodontics*, 1856-1859.

Salem Milani, A., Rahimi, S., Froughreyhani, M. and Vahid Pakdel, M. (2013). Effect of blood contamination on marginal adaptation and surface microstructure of Mineral trioxide aggregate: A SEM study. *Journal of Dental Research, Dental Clinics, Dental Prospects*, 157-163.

Samiei, M., Shahi, S., Aslaminabadi, N., Valizadeh, H., Aghazadeh, Z. and Pakdel, S. M. (2015). A new simulated plasma for assessing the solubility of mineral trioxide aggregate. *Iranian Endodontic Journal*, 30-34.

Septodont. Biodentine™: Active Biosilicate Technology™. *Septodont, Research and Development Department*, Paris, 2010. Available from: https://biodentine.com/wp-content/uploads/2020/08/Biodentine-Scientific-File.pdf.

Sharma, H. (2014). Evaluation of effect of 17% EDTA and 5.25% sodium hypochloride irrigating solutions on surface hardness of Brasseler Endosequence Root Repair Material [Thesis, Marquette University.]

Sheykhrezae, M. S., Meraji, N., Ghanbari, F., Nekoofar, M. H., Bolhari, B. and Dummer, P. M. H. (2018). Effect of blood contamination on the compressive strength of three calcium silicate-based cements. *Australian Endodontic Journal*, 255-259.

Shokouhinejad, N., Razmi, H., Khoshkhounejad, M., Firouzmanesh, P. (2017). The effect of blood and synthetic tissue fluid on the microhardness of ProRoot MTA, OrthoMTA and RetroMTA. *Journal of Dental Medicine Tehran University of Medical Sciences*, 136-143.

Singh, S., Podar, R., Dadu, S., Kulkarni, G., Vivrekar, S. and Babel, S. (2016). An in vitro comparison of push-out bond strength of biodentine and mineral trioxide aggregate in the presence of sodium hypochlorite and chlorhexidine gluconate. *Endodontology*, 42-45.

Smith, J. B., Loushine, R. J., Weller, R. N., Rueggeberg, F. A., Whitford, G. M., Pashley, D. H. and Tay, F. R. (2007). Metrologic evaluation of the surface of white MTA after the use of two endodontic irrigants. *Journal of Endodontics*, 463-467.

Song, M., Yue, W., Kim, S., Kim, W., Kim, Y., Kim, J. W. and Kim, E. (2016). The effect of human blood on the setting and surface micro-hardness of calcium silicate cements. *Clinical Oral Investigations*, 1997-2005.

Susumu, S., Keisuke, H., Masafumi, K., Yoshio, Y. and Masahiro, S. (2020). Effect of antiseptic agents on the surface microhardness of calcium silicate-based materials. *Biomedical Journal of Csientific & Technical Research*, 21350-21356.

Tingey, M. C., Bush, P. and Levine, M. S. (2008). Analysis of mineral trioxide aggregate surface when set in the presence of fetal bovine serum. *Journal of Endodontics*, 45-49.

Topçuoğlu, H. S., Arslan, H., Akçay, M., Saygili, G., Çakici, F. and Topçuoğlu, G. (2014). The effect of medicaments used in endodontic regeneration technique on the dislocation resistance of mineral trioxide aggregate to root canal dentin. *Journal of Endodontics*, 2041-2044.

Torabinejad, M., Higa, R. K., McKendry, D. J. and Pitt Ford, T. R. (1994). Dye leakage of four root end filling materials: effects of blood contamination. *Journal of Endodontics*, 159-163.

Turk, T., Ozisik, B. and Aydin, B. (2015). Time-dependent effectiveness of the intracanal medicaments used for pulp revascularization on the dislocation resistance of MTA. *BMC Oral Healt,* 130.

Üstün, Y., Topçuoğlu, H. S., Akpek, F. and Aslan, T. (2015). The effect of blood contamination on dislocation resistance of different endodontic reparative materials. *Journal of Oral Science*, 185-190.

Yavari, H. R., Borna, Z., Rahimi, S., Shahi, S., Valizadeh, H. and Ghojazadeh, M. (2013). Placement in an acidic environment increase the solubility of white mineral trioxide aggregate. *Journal of Conservative Dentistry*, 257-260.

Yavuz, H. F. and Eyüboğlu, G. B. (2021). The effect of blood contamination on shear bond strength of calcium silicate-based pulp capping materials. *Cumhuriyet Dental Journal*, 371-385.

Žižka, R., Čtvrtlík, R., Tomaštík, J., Fačevicová, K., Gregor, L. and Šedý, J. (2018). In vitro mechanical properties of mineral trioxide aggregate in moist and dry intracanal environments. *Iranian Endodontic Journal*, 20-24.

Chapter 5

Bibliography

10th International Symposium on High-Temperature Metallurgical Processing

LCCN	2019763626
Type of material	Book
Main title	10th International Symposium on High-Temperature Metallurgical Processing / edited by Tao Jiang, Jiann-Yang Hwang, Dean Gregurek, Zhiwei Peng, Jerome P. Downey, Baojun Zhao, Onuralp Yücel, Ender Keskinkilic, Rafael Padilla.
Edition	1st ed. 2019.
Published/Produced	Cham: Springer International Publishing: Imprint: Springer, 2019.
Description	1 online resource (XXI, 817 pages 475 illustrations, 350 illustrations in color.) text file PDF
ISBN	9783030059552
Related names	Downey, Jerome P, editor. Gregurek, Dean, editor. Hwang, Jiann-Yang, editor. Jiang, Tao, editor. Keskinkilic, Ender, editor. Padilla, Rafael, editor. Peng, Zhiwei, editor. Yücel, Onuralp, editor. Zhao, Baojun, editor.
Summary	In recent years, global metallurgical industries have experienced fast and prosperous growth. High-temperature metallurgical technology is the backbone to support the technical, environmental, and economical needs for this growth. This collection features contributions covering the advancements and developments of new high-

temperature metallurgical technologies and their applications to the areas of processing of minerals; extraction of metals; preparation of refractory and ceramic materials; sintering and synthesis of fine particles; treatment and recycling of slag and wastes; and saving of energy and protection of environment. The volume will have a broad impact on the academics and professionals serving the metallurgical industries around the world.

Contents Part 1. Energy-efficient and Clean Metallurgical Technology -- Simplified Process for Making Anode Copper -- Techno-economic Analysis of Energy Recovery from Plastic Waste -- Development of Continuous Blast Furnace Slag Solidification Process for Coarse Aggregates -- An Innovative Oxygen-enriched Flash Smelting Technology for Lead Smelting and its Industrial Application -- Characteristics and Control Technology of Fine Particulate Matter (PM) of Iron Ore Sintering -- Sintering Bed Spraying Steam to Reduce NOx and Dioxin Emissions in Shougang -- Part 2. Simulation and Modeling of High Temperature Metallurgical Process -- Neural Prediction Model for Extraction of Germanium from Zinc Oxide Dust by Microwave Alkaline Roasting-Water Leaching -- Simulation of Velocity Field of Molten Steel in Electric Arc Furnace Steelmaking -- Thermodynamic Modelling of Magnesium, Calcium and Strontium-oxides Reduction Systems in Vacuum -- Metallization and Carburization Kinetics in DR Shaft Furnaces. The Metcarb Model -- CFD Modeling of Flow and Chemical Reactions in a Submerged Lance Copper Smelting Furnace -- Numerical Simulation of Ultrasound-Induced Cavitation Bubbling in a Calcium Ferrite Melt -- Part 3. Alloys and Materials Preparation -- Synthesis of Nanocrystalline Carbide Ceramics via Reduction of Anion-loaded Activated Carbon Precursors -- Production of Lithium-Ion

Cathode Material for Automotive Batteries Using Melting Casting Process -- Part 4. Fundamental Research on High Temperature Metallurgical Processing -- Degradation Mechanisms of Refractories in a Bottom Blown Copper Smelting Furnace -- Reaction Routes of CaO-Fe2O3-TiO2 and Calcium Ferrite-TiO2 System in Continuous Heating Process -- Thermodynamic Calculations on Electric Furnace Smelting Separation of Chromium-bearing Vanadium Titanium Magnetite -- Preparation for High Activity Lime and its Effect on Desulfurization of Hot Metal Pretreatment -- Part 5. Extraction and Recovery of Metals -- An Industry Overlook of Secondary Lead Pyrometallurgical Processing -- Recovery of Aluminium and its Compounds with Hydro and Pyrometalurgical Methods from Non-metallic Residue -- Purification of Molten Zinc Chloride-Alkali Chloride by Cementation Reaction -- Thermodynamic Analysis of Smelting of Spent Catalysts for Recovery of Platinum Group Metals -- Preparation of Titanium Foams through Direct Electrolysis of the Sintered CaO-TiO2 in Molten Salt CaCl2 -- Experimental Study on Oxidative Desulfurization and Selective Reduction of Molten Copper Slag -- Recycling SiO2 and Al2O3 from the Metallurgical Slag of Nickel Laterite Ores in Molten Sodium Hydroxides -- Remove Sulfur in Copper Dross from Refining Lead by Converting Process -- Part 6. Treatment and Recycling of Metallurgical Slag/Solid Wastes -- Recovery of Fe-Cu Alloys from Copper Slags -- Physiochemical Properties of High Alumina Blast Furnace Slag -- Effect of Cooling Rate on the Acidolysis of Titania Slag -- Structural Analysis of Ge-containing Ferrous Calcium Silicate Magnesia Slag for Applications of Black Copper Smelting -- Selective Recovery of P and Mn from Steelmaking Slag by Carbothermic Reduction -- The Use of Zirconia-based Solid Electrolytes Oxygen Sensor in High Titanium Slag -- In-situ Observation of the Precipitation Behavior of Dy2O3 Containing Slag System -- Recovery of Zn and Mn from Spent Alkaline Batteries -- Part 7. Ironmaking, Steelmaking and Casting -- Optimization of Exothermic Riser Sleeve Design Parameters -- Assessment of Gas-Slag-Metal Interaction during a Converter Steelmaking Process -- On the Role of Nb on the Texture and Mechanical Properties of a Novel As-rolled Medium Carbon Wear Resistant Slurry Pipeline Steel -- A Statistical Analysis of Process Abnormalities in Slab Casting -- Effect of Density Difference on Particle Segregation Behaviors at Bell-less Top Blast Furnace with Parallel-type Hopper -- The Effect of Austenitizing Temperature on Hardenability, Precipitation and Mechanical Properties of Boron Bearing Cr-Mo Alloy Steel -- Part 8. Agglomeration and Direct Reduction of Complex Iron Ores -- Study on Direct Reduction Melting Separation-leaching Process of Disposal Rare Earth Composite Iron Ore -- Reduction Behavior of Garnierite Using Methane by Roasting-Magnetic Separation Method -- Effect of Calculation Method of CaO Addition on Liquid Phase Fluidity -- Effect of Carbon Coating on Magnetite Reduction -- Optimization Method for Iron Ore Blending Based on the Sintering Basic Characteristics of Blended Ore -- Study on Direct Reduction of Low-grade Iron Ore-coal Mini-pellets in Coal-based Rotary Kiln -- Part 9. Poster Session I -- Controlled Synthesis of TiC Nanoparticles Using Solid Oxide Membrane Technology in Molten CaCl2 -- Effect of Chemical Components of Mould Flux on Dissolution Rate of Al2O3 into Molten Flux for High Manganese High Aluminum Steel -- Effect of Temperature on Oxidation Behavior of Cr-Mo-V Steel with Different Cr Contents for High-speed Train Brake Discs --

Electrochemical Preparation of Ti5Si3/TiC Composite from Titanium-rich Slag in Molten CaCl2 -- Evolution of Al-Ti-Mg-O Inclusions during Refining and Casting Process of Interstitial Free Steel -- Experimental Study on Carburization of Higher Vanadium-bearing Hot Metal -- Hematite Precipitation from High Iron Solution in Hydrometallurgy Process -- Influence On The Crystallization Phase Of Mold Flux By Magnetic Fields -- Kinetics Study on Limestone Decomposition in Early Converter Slag -- Mathematical Modeling and Analysis of Converter Slagging and Steelmaking Process by Replacing Part of Lime With Limestone -- Research of Digital Platform and Process Guidance Model in EAF Steelmaking Process -- Research on Factors Affecting and Prediction Model of Silicon Content in Hot Metal of Corex -- Studied on the Cooling Effect of CO2 on the Temperature of Vanadium in Converter -- Study on Grain Size and Porosity of the Produced Lime from Limestone in Early Converter Slag -- Study on Reducing Al2O3 Inclusions by Optimizing Refining Slag -- Study on the Volatilization of Sb2S3 in Vacuum -- The Effects of ZrO2, Y2O3 and Sc2O3 on the Properties of Mould Fluxes for High Manganese High Aluminum Steels -- Thermogravimetric Analysis and Kinetic Study of the Calcification Roasting of Vanadium Slag -- Viscosity of Mould Flux under Electromagnetic Field -- Part 10. Poster Session II -- Analysis of Microwave Drying Behavior of Nickel Laterite -- Analysis of Operational Parameters Affecting Metallization Degree of DRI in Reduction Shaft of COREX Process and Improvement Measures -- Dechlorination of Zinc Oxide Dust by Microwave Rosating with RSM Optimization -- Effect of TiO2 on the Viscous Behavior of High Alumina Blast Furnace Slag -- Fundamental Research on the Iron Nugget Process

	from Carbon Composite Pellet -- Influence of Coke Quality on Main Technical Indexes of Blast Furnace -- Kinetic Analysis of Blast Furnace Dust Recycling with Flash Reduction Process at High Temperature -- Preparation and Characterization of Iron-coke Briquette -- Preparation of Oxidized pellets with Chrome Ore -- Research and Application of Sintering Surface Steam Spraying Technology for Energy Saving and Quality Improvement -- Research on Bonding Mechanism of Sintering Grate -- Research on Optimizing Sinter Ore Matching Based on the High Temperature Characteristic Numbers -- Research on the Mineral Composition and Microstructure Changes of Iron Ore Sinter during the Gas-Solid Reduction -- Roasting Kinetics of Molybdenite Concentrates -- Study on Influences of Different Ti-bearing Materials on MgO-bearing Pellets Metallurgical Properties -- Supergravity Separation of Pb and Sn from Waste Printed Circuit Boards -- The Effect of Temperature and Additive on Transport and Transformation of P of High-phosphorus Iron Ore during Carbothermic Reduction -- Thermodynamic Calculations on Direct Reduction of Chromium-bearing Vanadium Titanium Magnetite.
LC Subjects	Engineering--Materials. Materials science. Metals.
Other Subjects	Characterization and Evaluation of Materials. Materials Engineering. Metallic Materials.
Additional formats	Print version: 10th International Symposium on High-Temperature Metallurgical Processing. 9783030059545 (DLC) 2018964681 Printed edition: 9783030059545 Printed edition: 9783030059569
Series	The Minerals, Metals & Materials Series, 2367-1181

The Minerals, Metals & Materials Series, 2367-1181

2nd International Symposium on Autoclaved Calcium Silicate Building Materials, Hannover 25/3 - 24-69; participants report,

LCCN	73983517
Type of material	Book
Personal name	Halvorsen, Ulf A.
Main title	2nd International Symposium on Autoclaved Calcium Silicate Building Materials, Hannover 25/3 - 2/4-69; participants report, by Ulf A. Halvorsen and Bekele W. Medhin.
Published/Created	Addis Ababa, Ethio-Swedish Institute of Building Technology, 1969.
Description	13, [11] l. 30 cm.
LC classification	TA455.S46 H34
Related names	Bekele W. Medhin, joint author. International Symposium on Autoclaved Calcium Silicate Building Materials, 2d, Hanover, 1969.
LC Subjects	Calcium silicates--Congresses.

9th International Symposium on High-Temperature Metallurgical Processing

LCCN	2019753136
Type of material	Book
Main title	9th International Symposium on High-Temperature Metallurgical Processing / edited by Jiann-Yang Hwang, Tao Jiang, Mark William Kennedy, Dean Gregurek, Shijie Wang, Baojun Zhao, Onuralp Yücel, Ender Keskinkilic, Jerome P Downey, Zhiwei Peng, Rafael. Padilla.
Edition	1st ed. 2018.
Published/Produced	Cham: Springer International Publishing: Imprint: Springer, 2018.
Description	1 online resource (XXIV, 887 pages 534 illustrations) text file PDF
ISBN	9783319721385
Related names	Downey, Jerome P. editor.

Gregurek, Dean. editor.
Hwang, Jiann-Yang. editor.
Jiang, Tao. editor.
Kennedy, Mark William. editor.
Keskinkilic, Ender. editor.
Padilla, Rafael. editor.
Peng, Zhiwei. editor.
Wang, Shijie. editor.
Yücel, Onuralp. editor.
Zhao, Baojun. editor.

Summary In recent years, global metallurgical industries have experienced fast and prosperous growth. High-temperature metallurgical technology is the backbone to support the technical, environmental, and economical needs for the growth. This collection features contributions covering the advancements and developments of new high-temperature metallurgical technologies and their applications to the areas of processing of minerals; extraction of metals; preparation of metallic, refractory and ceramic materials; treatment and recycling of slag and wastes; and saving of energy and protection of environment.The volume will have a broad impact on the academics and professionals serving the metallurgical industries around the world.

Contents Part 1. Energy-efficient and Clean Metallurgical Technology -- Simplified Process for Making Anode Copper -- Techno-economic Analysis of Energy Recovery from Plastic Waste -- Development of Continuous Blast Furnace Slag Solidification Process for Coarse Aggregates -- An Innovative Oxygen-enriched Flash Smelting Technology for Lead Smelting and its Industrial Application -- Characteristics and Control Technology of Fine Particulate Matter (PM) of Iron Ore Sintering -- Sintering Bed Spraying Steam to Reduce NOx and Dioxin Emissions in Shougang -- Part 2. Simulation and Modeling of High

Temperature Metallurgical Process -- Neural Prediction Model for Extraction of Germanium from Zinc Oxide Dust by Microwave Alkaline Roasting-Water Leaching -- Simulation of Velocity Field of Molten Steel in Electric Arc Furnace Steelmaking -- Thermodynamic Modelling of Magnesium, Calcium and Strontium-oxides Reduction Systems in Vacuum -- Metallization and Carburization Kinetics in DR Shaft Furnaces. The Metcarb Model -- CFD Modeling of Flow and Chemical Reactions in a Submerged Lance Copper Smelting Furnace -- Numerical Simulation of Ultrasound-Induced Cavitation Bubbling in a Calcium Ferrite Melt -- Part 3. Alloys and Materials Preparation -- Synthesis of Nanocrystalline Carbide Ceramics via Reduction of Anion-loaded Activated Carbon Precursors -- Production of Lithium-Ion Cathode Material for Automotive Batteries Using Melting Casting Process -- Part 4. Fundamental Research on High Temperature Metallurgical Processing -- Degradation Mechanisms of Refractories in a Bottom Blown Copper Smelting Furnace -- Reaction Routes of CaO-Fe2O3-TiO2 and Calcium Ferrite-TiO2 System in Continuous Heating Process -- Thermodynamic Calculations on Electric Furnace Smelting Separation of Chromium-bearing Vanadium Titanium Magnetite -- Preparation for High Activity Lime and its Effect on Desulfurization of Hot Metal Pretreatment -- Part 5. Extraction and Recovery of Metals -- An Industry Overlook of Secondary Lead Pyrometallurgical Processing -- Recovery of Aluminium and its Compounds with Hydro and Pyrometalurgical Methods from Non-metallic Residue -- Purification of Molten Zinc Chloride-Alkali Chloride by Cementation Reaction -- Thermodynamic Analysis of Smelting of Spent Catalysts for Recovery of Platinum Group Metals -- Preparation of Titanium Foams through Direct

Electrolysis of the Sintered CaO-TiO2 in Molten Salt CaCl2 -- Experimental Study on Oxidative Desulfurization and Selective Reduction of Molten Copper Slag -- Recycling SiO2 and Al2O3 from the Metallurgical Slag of Nickel Laterite Ores in Molten Sodium Hydroxides -- Remove Sulfur in Copper Dross from Refining Lead by Converting Process -- Part 6. Treatment and Recycling of Metallurgical Slag/Solid Wastes -- Recovery of Fe-Cu Alloys from Copper Slags -- Physiochemical Properties of High Alumina Blast Furnace Slag -- Effect of Cooling Rate on the Acidolysis of Titania Slag -- Structural Analysis of Ge-containing Ferrous Calcium Silicate Magnesia Slag for Applications of Black Copper Smelting -- Selective Recovery of P and Mn from Steelmaking Slag by Carbothermic Reduction -- The Use of Zirconia-based Solid Electrolytes Oxygen Sensor in High Titanium Slag -- In-situ Observation of the Precipitation Behavior of Dy2O3 Containing Slag System -- Recovery of Zn and Mn from Spent Alkaline Batteries -- Part 7. Ironmaking, Steelmaking and Casting -- Optimization of Exothermic Riser Sleeve Design Parameters -- Assessment of Gas-Slag-Metal Interaction during a Converter Steelmaking Process -- On the Role of Nb on the Texture and Mechanical Properties of a Novel As-rolled Medium Carbon Wear Resistant Slurry Pipeline Steel -- A Statistical Analysis of Process Abnormalities in Slab Casting -- Effect of Density Difference on Particle Segregation Behaviors at Bell-less Top Blast Furnace with Parallel-type Hopper -- The Effect of Austenitizing Temperature on Hardenability, Precipitation and Mechanical Properties of Boron Bearing Cr-Mo Alloy Steel -- Part 8. Agglomeration and Direct Reduction of Complex Iron Ores -- Study on Direct Reduction Melting Separation-leaching Process of Disposal Rare Earth Composite Iron Ore --

Reduction Behavior of Garnierite Using Methane by Roasting-Magnetic Separation Method -- Effect of Calculation Method of CaO Addition on Liquid Phase Fluidity -- Effect of Carbon Coating on Magnetite Reduction -- Optimization Method for Iron Ore Blending Based on the Sintering Basic Characteristics of Blended Ore -- Study on Direct Reduction of Low-grade Iron Ore-coal Mini-pellets in Coal-based Rotary Kiln -- Part 9. Poster Session I -- Controlled Synthesis of TiC Nanoparticles Using Solid Oxide Membrane Technology in Molten CaCl2 -- Effect of Chemical Components of Mould Flux on Dissolution Rate of Al2O3 into Molten Flux for High Manganese High Aluminum Steel -- Effect of Temperature on Oxidation Behavior of Cr-Mo-V Steel with Different Cr Contents for High-speed Train Brake Discs -- Electrochemical Preparation of Ti5Si3/TiC Composite from Titanium-rich Slag in Molten CaCl2 -- Evolution of Al-Ti-Mg-O Inclusions during Refining and Casting Process of Interstitial Free Steel -- Experimental Study on Carburization of Higher Vanadium-bearing Hot Metal -- Hematite Precipitation from High Iron Solution in Hydrometallurgy Process -- Influence On The Crystallization Phase Of Mold Flux By Magnetic Fields -- Kinetics Study on Limestone Decomposition in Early Converter Slag -- Mathematical Modeling and Analysis of Converter Slagging and Steelmaking Process by Replacing Part of Lime With Limestone -- Research of Digital Platform and Process Guidance Model in EAF Steelmaking Process -- Research on Factors Affecting and Prediction Model of Silicon Content in Hot Metal of Corex -- Studied on the Cooling Effect of CO2 on the Temperature of Vanadium in Converter -- Study on Grain Size and Porosity of the Produced Lime from Limestone in Early Converter Slag -- Study on Reducing Al2O3 Inclusions by

Optimizing Refining Slag -- Study on the Volatilization of Sb2S3 in Vacuum -- The Effects of ZrO2, Y2O3 and Sc2O3 on the Properties of Mould Fluxes for High Manganese High Aluminum Steels -- Thermogravimetric Analysis and Kinetic Study of the Calcification Roasting of Vanadium Slag -- Viscosity of Mould Flux under Electromagnetic Field -- Part 10. Poster Session II -- Analysis of Microwave Drying Behavior of Nickel Laterite -- Analysis of Operational Parameters Affecting Metallization Degree of DRI in Reduction Shaft of COREX Process and Improvement Measures -- Dechlorination of Zinc Oxide Dust by Microwave Rosating with RSM Optimization -- Effect of TiO2 on the Viscous Behavior of High Alumina Blast Furnace Slag -- Fundamental Research on the Iron Nugget Process from Carbon Composite Pellet -- Influence of Coke Quality on Main Technical Indexes of Blast Furnace -- Kinetic Analysis of Blast Furnace Dust Recycling with Flash Reduction Process at High Temperature -- Preparation and Characterization of Iron-coke Briquette -- Preparation of Oxidized pellets with Chrome Ore -- Research and Application of Sintering Surface Steam Spraying Technology for Energy Saving and Quality Improvement -- Research on Bonding Mechanism of Sintering Grate -- Research on Optimizing Sinter Ore Matching Based on the High Temperature Characteristic Numbers -- Research on the Mineral Composition and Microstructure Changes of Iron Ore Sinter during the Gas-Solid Reduction -- Roasting Kinetics of Molybdenite Concentrates -- Study on Influences of Different Ti-bearing Materials on MgO-bearing Pellets Metallurgical Properties -- Supergravity Separation of Pb and Sn from Waste Printed Circuit Boards -- The Effect of Temperature and Additive on Transport and Transformation of P of High-phosphorus Iron Ore

	during Carbothermic Reduction -- Thermodynamic Calculations on Direct Reduction of Chromium-bearing Vanadium Titanium Magnetite.
LC Subjects	Engineering--Materials. Materials science. Metals.
Other Subjects	Characterization and Evaluation of Materials. Materials Engineering. Metallic Materials.
Additional formats	Print version: 9th International Symposium on High-Temperature Metallurgical Processing. 9783319721378 (DLC) 2017959922 Printed edition: 9783319721378 Printed edition: 9783319721392 Printed edition: 9783319891477
Series	The Minerals, Metals & Materials Series, 2367-1181 The Minerals, Metals & Materials Series, 2367-1181

Autoclaved calcium silicate building products:

LCCN	70415314
Type of material	Book
Main title	Autoclaved calcium silicate building products: papers read at the symposium held at the School of Pharmacy (University of London), 18th-21st May, 1965, with the discussions that followed.
Published/Created	London, Society of Chemical Industry, 1967.
Description	v, 277 p. illus., tables, diagrs. 28 1/2 cm.
LC classification	TP785 .A84 1967
Related names	University of London. School of Pharmacy. Society of Chemical Industry (Great Britain)
LC Subjects	Calcium silicates--Congresses. Building materials--Congresses.
Notes	Includes bibliographies.

Autoclaved calcium silicate building products; papers read at the Symposium held at the School of Pharmacy (University of London), 18th-21st May, 1965, with the discussions that followed.

LCCN	68093562
Type of material	Book
Main title	Autoclaved calcium silicate building products; papers read at the Symposium held at the School of Pharmacy (University of London), 18th-21st May, 1965, with the discussions that followed.
Published/Created	London, Society of Chemical Industry; New York, Gordon & Breach, Science Publications, 1967.
Description	277 p. illus. 29 cm.
LC classification	TP785 .A84
Related names	Society of Chemical Industry (Great Britain) University of London. School of Pharmacy.
LC Subjects	Calcium silicates--Congresses. Building materials--Congresses.
Notes	Includes bibliographies.

British standard specification for sand-lime (calcium-silicate) bricks.

LCCN	23014166
Type of material	Book
Corporate name	British Engineering Standards Association.
Main title	British standard specification for sand-lime (calcium-silicate) bricks.
Published/Created	London, Pub. for the association by C. Lockwood & son, 1923.
Description	11 p. 22 cm.
LC classification	TP832.S2 B7
LC Subjects	Sand-lime brick.

Eco-materials processing and design XIV: selected, peer reviewed papers from the 14th International Symposium on Eco-materials Processing & Design (ISEPD), January 15-18, 2013, Kagoshima, Japan

LCCN	2013427378
Type of material	Book
Meeting name	International Symposium on Eco-Materials Processing & Design (14th: 2013: Kagoshima, Japan)

Main title Eco-materials processing and design XIV: selected, peer reviewed papers from the 14th International Symposium on Eco-materials Processing & Design (ISEPD), January 15-18, 2013, Kagoshima, Japan / edited by T. Kai ... [et al.].

Published/Created Durnten-Zurich, Switzerland; Enfield, NH: Trans Tech Publications, 2013.

Description xi, 174 p.: ill.; 24 cm.

ISBN 9783037857144
3037857145

LC classification TD170 .I57 2013

Portion of title 14th International Symposium on Eco-Materials Processing and Design
ISEPD2013
Ecomaterials processing and design

Related names Kai, Takami.

Summary The International Symposium on Eco-materials Processing and Design, conjugated with the China-Japan-Korea A3 Foresight Program Seminar is a series of conferences that provide the presentations, exchange of information and discussion on state-of-the art of scientific and industrial development of environmentally friendly materials and systems. Recent symposia were organized in Japan (2010), Thailand (2011) and China (2012). The 14th ISEPD was held at Kagoshima, Japan, on January 15-18, 2013.Researchers, engineers and students enjoyed the academic program of the exciting symposium in beautiful Kagoshima, and to exchange and share the scientific knowledge. ISEPD 2013 covered the following five topics: (A) Materials for energy conversion, solar cell and new energy, (B) Photocatalysis and semiconducting materials, (C) Processing and design for eco-materials, (D) Bio-inspired materials and hybrid materials, and (E) Porous and nanostructured materials.

Contents Synthesis of polyglycerol fatty acid ester for insulating oil / Miseong Gwon ... [et al.] -- Preparation of the catalyst for methanol steam reforming from Cr-Zr amorphous alloys / Toshiya Matsusako ... [et al.] -- Evaluation of electric power of SOFC using reformed biogas / Naoki Furukawa ... [et al.] -- Optical and photovoltaic properties of CdS/Ag_2S quantum dots co-sensitized-solar cells / Auttasit Tubtimtae, Ming Way Lee -- Preparation and electrical conductivity of ... perovskite solid solution (x = 0-0.5) / Ryuji Oitate ... [et al.] -- The thermal expansion characteristics of epoxy resin/crepe paper composites for high-voltage DC bushing / Naikui Gao ... [et al.] -- Fabrication and thermoelectric properties of Al-Doped $(ZnO)_5In_2O_3$ by microwave heating / Tomoyuki Endo ... [et al.] -- Nitrogen doping to metatitanic acid by NH_3 heat treatment for achieving visible-light-responsive photocatalytic activity / Se Keun Park, Jun Ho Eun, Hyun Ho Shin -- Feasibility of hydrothermal nitrogen doping process for enhancing visible light photocatalytic activity of metatitanic acid power / Se Keun Park, Ji Eun Woo, Hyun Ho Shin -- Preparation of hematite by drip thermal oxidation using a fluidized bed for conversion to electromagnetic-wave-absorptive magnetite / Tsutomu Nakazato, Yoichi Iikuma, Takami Kai -- Fabrication and consolidation of TiN_x by pulsed electric current sintering / Santiago I. Suarez Vazquez, Makoto Nanko -- Fabrication and characterization of calcium silicate phosphors -Ca_2SiO_4 and $Ca_2MgSi_2O_7$- / Takanori Watari ... [et al.] -- Recovery of silicon and silicon carbide powder from waste silicon wafer sludge / Woo Teck Kwon ... [et al.] -- Characterization of silicon kerf and the effect of grinding on the synthesis of SiC powder / Woo Teck Kwon ... [et al.] -- A model for stress history during sintering of ceramic laminates / Kouichi Yasuda, Peiling Lv -- Simulation study on HPDC process for automobile part with aluminum alloy / Hong Kyu Kwon, Kwant Kyu Seo -- Effects of CNT addition on mechanical properties, electric conductivity and oxidation resistance of Al2O3-TiC composite / Hideaki Sano ... [et al.] -- Eco-fabrication of metal nanoparticle related materials by non-equilibrium reaction field / Yamoto Hayashi ... [et al.] -- Synthesis and characterization of zeolite from glass by hydrothermal processing / Masato Tsujiguchi ... [et al.] -- Enhancement of grain refinement and formability of cross roll rolled Ni-0Cr alloy / Kuk Hyun Song, Han Sol Kim, Won Yong Kim -- Effect of fiber cross-sectional shape and nano ZnO/modified chitosan loading on antibacterial properties / Natthaphop Suwannamek, Nanjaporn Ruangpaisarn, Chureerat Prahsan -- Linear assembly of oxidized surface treated nanodiamonds in polymer-based nanohybrids by electric field inducement / Son Thanh Nguyen ... [et al.] -- Characterization of HA/Ta composite coatings fabricated by vacuum plasma spraying / Li Ping Huang ... [et al.] -- Responses of human bone marrow-derived stromal cells to calcium silicate coatings with different chemical stability / You Tao Xie, Li Ping Huang, Xue Bin Zheng -- Preparation of copper nanoparticles by pulsed wire discharge in water / Yoshinori Tokoi ... [et al.] -- Oxidation of tantalum nitride / Kazuya Hamaguchi, Tomoyuki Tsuchiyama, Junichi Matsushita -- Fabrication of nanorod array structure by using aluminizing and internal oxidation of Ni with micro-channels / Takeshi Ishizaki, Dung Thi Mai Do, Makoto Nanko -- Characterization of hydroxyapatite containing of titanium hydride sintered body by nano size powder / Kazuki Ishiwata, Kazumasa Tawara, Junichi Matsushita -- Characterization and cell response of novel hydroxyapatite/glyceride-based polyurethane porous scaffold / Jing Jing Du ... [et al.] -- Synthesis of palladium nanoparticles in various carbon materials / Min A Bae, Jae Ho Baek, Man Sig Lee -- Compressive properties of porous aluminum alloy fabricated by joining pipes and melt through

	continuous casting / Junichi Ichikawa, Tatsuro Hayashida, Shinsuke Suzuki -- Effect of pore forming agent on microstructures and properties of porous molybdenum / Zhen Lin Lu, Xiao Jie Rao, Xiao Feng Xu -- Dispersion and impact strength of epoxy resin/nano-Al2O3 composites / Nai Kui Gao ... [et al.] -- Relationship between nitrogen content and mechanical properties in ... thin films / Thi Huyen Tram Pham ... [et al.].
LC Subjects	Materials management--Environmental aspects--Congresses. Materials--Environmental aspects--Congresses. Environmental engineering--Congresses. Environmental protection--Design--Congresses.
Notes	The 14th ISEPD was held at Kagoshima, Japan, on January 15-18, 2013."--Preface. Includes bibliographical references and indexes.
Series	Materials science forum, 0255-5476; v. 761 Materials science forum; v. 761.

Mineral trioxide aggregate: properties and clinical applications

LCCN	2014004603
Type of material	Book
Main title	Mineral trioxide aggregate: properties and clinical applications / edited by Dr. Mahmoud Torabinejad.
Published/Produced	Chichester, West Sussex, UK Ames, Iowa: Wiley, Blackwell, 2014.
Description	1 online resource.
ISBN	9781118892503 (ePub) 9781118892466 (Adobe PDF)
LC classification	RK351
Related names	Torabinejad, Mahmoud, editor of compilation.
Contents	Pulp and periradicular pathways, pathosis, and closure / Mahmoud Torabinejad -- Chemical properties of mineral trioxide aggregate / David W. Berzins -- Physical properties of mineral trioxide aggregate / Lawrence Gettleman and Ricardo Caicedo -- Use of MTA in vital pulp therapy / Till Dammaschke, Joe Camp, and George Bogen -- Use

	of MTA in teeth with necrotic pulps and open apices / Shahrokh Shabahang and David Witherspoon -- Use of MTA in regenerative endodontics / Mahmoud Torabinejad, Robert P. Corr, and George T.J. Huang -- Use of MTA as root perforation repair / Mahmoud Torabinejad and Ron Lemon -- Root canal obturation with MTA / George Bogen, Ingrid Lawaty, and Nicholas Chandler -- Use of MTA as root end filling / Seung-Ho Baek and Su-Jong Shin -- Calcium silicate-based cements / Masoud Parirokh and Mahmoud Torabinejad.
Other Subjects	Root Canal Filling Materials. Aluminum Compounds. Biocompatible Materials. Calcium Compounds. Root Canal Therapy--methods. Silicates.
Notes	Includes bibliographical references and index.
Additional formats	Print version: Mineral trioxide aggregate Ames, Iowa: John Wiley & Sons Inc., 2014 9781118401286 (DLC) 2014003707

Phosphorus Recovery and Recycling

LCCN	2019751883
Type of material	Book
Main title	Phosphorus Recovery and Recycling / edited by Hisao Ohtake, Satoshi Tsuneda.
Edition	1st ed. 2019.
Published/Produced	Singapore: Springer Singapore: Imprint: Springer, 2019.
Description	1 online resource (XV, 526 pages 239 illustrations, 130 illustrations in color.) text file PDF
ISBN	9789811080319
Related names	Ohtake, Hisao. editor. Tsuneda, Satoshi. editor.
Summary	This book focuses on the engineering aspects of phosphorus (P) recovery and recycling, presenting recent research advances and applications of

technologies in this important and challenging area of engineering. It highlights full-scale applications to illustrate the performance and effectiveness of the new technologies. As an essential element for all living organisms, P cannot be replaced by any other element in biochemical processes, humans ultimately rely its availability. Today, P is mostly obtained from mined rock phosphate (Pi). However, natural reserves of high-grade rock Pi are limited and dwindling on a global scale. As such, there have been increased efforts to recycle P from secondary sources, including sewage sludge, animal manure, food waste, and steelmaking slag, and so close the anthropogenic P cycle. In addition to various aspects of phosphorus covered by other literature,including chemistry, biochemistry, ecology, soil-plant systems and sustainable management, this book is a valuable and comprehensive source of information on the rapidly evolving field of P recovery and recycling engineering for students, researchers, and professionals responsible for sustainable use of phosphorus.

Contents

Foreword -- Preface -- Part I Phosphorus Flow and Recycling -- 1 Development of Phosphorus Recycling in Europe and Japan -- 2 Phosphorus Flows in Asia -- 3 Circular Economy - Bridging the Gap between Phosphorus Recovery and Recycling -- 4 Life Cycle Assessment of Processes for Phosphorus Recycling -- 5 Phosphorus: Reserves, Production and Applications -- 6 Success Factors for Implementing Phosphorus Recycling Technologies -- Part II Incinerated Sludge Ash -- 7 Industrial-Scale Manufacturing of Phosphoric Acid Using Sewage Sludge Ash -- 8 Alkaline Leaching of Phosphate from Sewage Sludge Ash -- 9 Phosphorus Recovery from Sewage Sludge Ash - A Case Study in Gifu, Japan -- 10 Urban Phosphorus-Mining in the Canton Zurich: Phosphoric Acid from

Sewage Sludge Ash -- 11 Calcination Technology for Manufacturing Mineral Fertilizer Using CaO-Enriched Sewage Sludge Ash -- 12 Phosphorus Recovery from Sewage Sludge by High-Temperature Thermochemical Process (KUBOTA process) -- 13 Phosphorus Extraction from Sludge Ash by the CO2 blowing method -- 14 Ecophos Process -- 15 Outotec (AshDec®) Process for Phosphorus-Fertilizer from Sewage Sludge Ash -- 16 Phosphorus-Recovery into Fertilizers and Industrial Products by ICL in Europe -- Part III Sewage Sludge and Night Soil -- 17 Struvite Recovery from Digested Sewage Sludge -- 18 Phosphorus Recovery from Night Soil and Johkasou Sludge -- 19 The Stuttgart Process (Germany) -- 20 Phosphorus Recovery from Wet Sewage Sludge Using CO2 -- 21 Effect of Iron on Phosphorus Recovery from Sewage Sludge -- Part IV Steelmaking Slag -- 22 Phosphorus Separation and Recovery from Steelmaking Slag -- 23 Extraction of Phosphorous from Dephosphorization Slag -- 24 Phosphorus in Steelmaking Process -- Part V Animal Manure -- 25 Energy-Effective Carbonization Technology -- 26 Hydrothermal Process for Extracting Phosphorus from Animal Manure -- 27 Recovery of Calcium Phosphate from Composted Chicken Manure and Industrial Waste -- 28 Outotec Manure, Slurry and Sludge Processing Technology -- 29 Bone Char as a Novel Phosphorus Fertilizer -- Part VI Solid Adsorbents -- 30 Phosphorus Recovery Using Amorphous Calcium Silicate Hydrates -- 31 High-Performance Phosphorus Adsorbent Based on Concrete Sludge -- 32 Valorisation of Nutrients in Wastewaters Using Reactive Inorganic Sorbents -- 33 Phosphate Separation from Aqueous Solution Using a Chitosan-Based Biodegradable Ion Exchanger -- Part VII Biotechnology -- 34 Biological Phosphine Oxidation and Its Application to Phosphorus

	Recycling -- 35 HeatPhos Process for Recovering Phosphorus from Bio-P Sludge before Anaerobic Digestion.
LC Subjects	Biochemical engineering. Biotechnology. Environmental chemistry. Environmental engineering. Environmental management. Sustainable development. Water pollution.
Other Subjects	Sustainable Development. Biochemical Engineering. Environmental Chemistry. Environmental Engineering/Biotechnology. Environmental Management. Waste Water Technology / Water Pollution Control / Water Management / Aquatic Pollution.
Additional formats	Print version: Phosphorus recovery and recycling 9789811080302 (DLC) 2018942182 Printed edition: 9789811080302 Printed edition: 9789811080326 Printed edition: 9789811340444
Other standard no.	10.1007/978-981-10-8031-9

Proceedings of the 10th International Congress for Applied Mineralogy (ICAM)

LCCN	2017465368
Type of material	Book
Meeting name	International Congress on Applied Mineralogy in the Minerals Industry (10th: 2011: Trondheim, Norway)
Main title	Proceedings of the 10th International Congress for Applied Mineralogy (ICAM) / Maarten A.T.M. Broekmans, editor. monograph.
Published/Created	Heidelberg; New York: Springer, ©2012.
Description	xxv, 823 pages: illustrations (some color), maps (some color); 29 cm
ISBN	9783642276828 (electronic bk.) 3642276822 (electronic bk.)

	9783642276811
	3642276814
LC classification	QE351 .I57 2011
Portion of title	ICAM
Related names	Broekmans, Maarten A. T. M., 1963-
Summary	This book comprises 96 peer-reviewed contributions submitted to the 10th ICAM Congress, held in Trondheim, Norway on 01-05 August 2011. Themes covered include: 1) Advanced materials, including high-performance technical ceramics and glasses, 2) Analytical techniques, instrumentation and automation, 3) Bio-mimetic mineral materials, medical mineralogy, 4) Construction materials including cement/SCMs, concrete, bricks, tiles, screeds, 5) Cultural heritage, stone artifacts and preservation, 6) Environment and energy mineralogy, including CO2 sequestration, 7) Geometallurgy and process mineralogy, and 8) Industrial minerals including gems, ore minerals, and mineral exploration.
Contents	Iodine K-Edge Exafs Spectroscopy of Iodine-Bearing AFm-(Cl2, CO3, SO4) / Laure Aimoz, Erich Wieland, Christine Taviot-Guého, Marika Vespa and Rainer Dähn -- Prediction of Mineral Processing Behavior of Bauxite Ores by XRD Cluster Analysis / Juliana Lívi Antoniassi, Henrique Kahn, Carina Ulsen and Maria Manuela M.L. Tassinari -- Metallogeny Of Culampajá Gold District, Argentina / Julio C. Avila and José E. Lazarte -- Systematic REE Partitioning in Cubo-Dodecahedral Fluorite from Belgium Revealed by Cathodoluminescence Spectral Imaging and Laser Ablation-ICP-MS / Jean-Marc Baele, Laurence Monin, Jacques Navez and Luc André -- Geochemical Characteristics of the Aligudarz Granite Intrusion and Its Mineralization Potential / Siyamak Bagheriyan -- Occurrences and Petrographical Variations among the Glauconite Deposits from Egypt / Hassan M. Baioumy, Sabah

N. Boulis and Mervat S. Hassan -- Investigation of an Old Problem: Optimisation and Cost Effectiveness / Robin N.J. Baldwin and Michael A. Eden -- Three-Dimensional Evaluation of Magnetite-Hematite Transformation Via EBSD Technique / Paola F. Barbosa, Leonardo E. Lagoeiro, Leonardo M. Graça and Ricardo A. Scholz -- Effect of Alteration on the Mineralogy and Flotation Performance of PPM Platinum Ore / Megan Becker, Mpho Ramonotsi and Jochen Petersen -- The Role of Process Mineralogy in Identifying the Cause of the Low Recovery of Chalcopyrite at KUCC / Dee J. Bradshaw, Brett Triffett and Dave Kashuba -- Petrographic Crystallography in Solving Problems of Applied Mineralogy / R.L. Brodskaya and E.L. Kotova -- Monomineralic Aggregates in Mortars: Examples From Lombardy (Northern Italy) / Roberto Bugini and Luisa Folli -- Assessment of Individual ASR-Aggregate Particles by XRD / Nélia Castro, Bjørn E. Sorensen and Maarten A.T.M. Broekmans -- Grain Size Analysis of Quartz in Potentially Alkali-Reactive Aggregates for Concrete: A Comparison Between Image Analysis and Point-Counting / Nélia Castro and Børge J. Wigum -- Effect of Farm Manure on Dissolution of Underlying Carbonate Rocks and Atmospheric CO2 Source/Sink / Liu Changli, Zhang Yun, Song Chao, Hou Hongbing and Jiang Jianmei, et al. -- High Voltage Selective Fragmentation for Detailed Mineralogical and Analytical Information, Case Study: Oiva's Gold-Quartz-Dyke in the Lapland Granulite Belt, Laanila, Northern Finland / Tegist Chernet.

The Use of 3D X-Ray Computed Tomography for Gold Location in Exploration Drill Cores / Deshenthree Chetty, Wilma Clark, Charles Bushell, Tlou Piet Sebola and Jakobus Hoffman, et al. -- Crandallite Formation in Archaeological Potteries Found in the Amazonian Dark Earth Soils /

Marcondes L. Costa, Suyanne F.S. Rodrigues, Glayce J.S. Silva and Herbert Pöllmann -- Bioleaching of Metals from Tail-Serpentine by Aspergillus niger / Faqin Dong, Yuan Tan, Wei Yang, Wei Zhang and Yuanyuan Zhou -- Chemical Composition and Age of Uraninite of the Zhovta Richka Uranium Deposit (Ukraine) / Alexander Emetz, Michel Cuney, Julien Mercadier and Yuriy Velikanov -- Metasomatic Formation and Replacement of Apatite (Bamble Sector, South Norway) / Ane K. Engvik, Håkon Austrheim, Peter Ihlen, Are Korneliussen and Andrew Putnis -- REE Mobilization in Complex Hydrothermal-Metasomatic Systems: Fluid Chemistry Evidence of Albitite And Chlorite-Talc Mineralisations in Central Sardinia, Italy / Sandro Fadda, Maddalena Fiori, Silvana Maria Grillo and Walter Prochaska -- Synthesis of Intermetallic Ni-Al Nanoparticles by Wet Chemistry Synthesis of Niacac2 and Alcl3 Precursors / Nor Akmal Fadil, Hideki Abe, Hideyuki Murakami, Saravanan Govindachetty and Kazuhiko Noda -- Building Façades of Oporto City Covered with Glazed Tiles. Guidelines of Haracterisation for Preservation / Luis Mariz Ferreira, João Coroado and Ana Luisa Velosa -- Advanced Discrimination of Hematite and Magnetite by Automated Mineralogy / German Figueroa, Kurt Moeller, Michael Buhot, Gerda Gloy and David Haberla -- The Discovery of a New Metallogenic Province in Sardinia (Italy): Au-Ag Mineralisations of Tertiary Age / Sandro Fadda, Maddalena Fiori and Silvana Maria Grillo -- The Formation of Oxalate Patina on the Surface of Carbonate Rocks Under the Influence of Microorganisms / Olga Frank-Kamemetskaya, Alexey Rusakov, Ekaterina Barinova, Marina Zelenskaya and Dmitrij Vlasov -- Paralava from Coal-Dump Combustion in Upper Silesia, Poland: Melt Separation Leaving a Cordierite-Rich Restite.

/ Aleksandra Gawęda, Mirosław Kierepka, Janusz Janeczek and Alicja Mazur-Cuber -- Vein Type Mineralization and Related Alterations of Narigun Polymetallic Deposit, East of Yazd, Central Iran / Mehdi Gharesi and Mehrdad Karimi -- The Roman Bridge of Sant'antioco (Sardinia, Italy): The Analysis of Masonries and Characterization of Materials / Caterina Giannattasio and Silvana Maria Grillo.
Traditional Masonry Techniques and Characterization of Materials: The Mezzaspiaggia Tower (Cagliari, Sardinia) / Caterina Giannattasio and Silvana Maria Grillo -- Morphological Characterization of Natural and Artificial Sands Through Image Analysis / Otávio da Fonseca Martins Gomes, Paulo Roberto Lopes Lima and Ariella Vianna Fontes -- Sem-Eds-Based Protocol for Subsurface Drilling Mineral Identification and Petrological Classification / David Haberlah, Michael Owen, Pieter W.S.K. Botha and Paul Gottlieb -- Petrological Reconstruction of the Subsurface Based on PDC Drill Cuttings: An Advanced Rock Typing Approach / David Haberlah, Pieter W.S.K. Botha, Nicole Dobrzinski, Alan R. Butcher and John G. Kaldi -- Reflected Light Microscopy of Industrial Minerals / Richard D. Hagni -- The Replacement of Carbonatite and Marble by Fluorspar Ores and Its Relationship to Beneficiation Problems at Okorusu, Namibia / Richard D. Hagni -- Ore Microscopic Study of Copper Losses from Mining of Bornite Ores in the Sweetwater Mine, Viburnum Trend, Southeast Missouri / Richard D. Hagni -- Integration and Analysis of Optical and Sem-Based Microscopy for Automated Mineralogical Characterisation / Richard Hartner, Steve G. Walters and Ron Berry -- Apatite-Biotite-Carbonatite (Stjernøy, N-Norway): Potential and Obstacles Regarding a Multinutrient Rock-Fertilizer / Michael Heim,

Maria H. Hillersøy, Marina A. Bleken, Håvard Gautneb and Elin L. Gjengedal -- Quantitative Mineral Characterization of Granitic Pegmatite Using Topas Rietveld XRD Refinement / Kristin Husebø Hestnes and Bjørn Eske Sørensen -- Formation of Calcium Oxalates in the Human Body / Alina R. Izatulina and Yuriy O. Punin -- Halloysite from the Eucla Basin, South Australia -- Comparison of Physical Properties for Potential New Uses / John L. Keeling, Pooria Pasbakhsh and G. Jock Churchman -- Geochemical Variations During Alteration of an Andesite-Basalt to Bentonite, In Khur, East of Isfahan, Iran / Mahmoud Khalili and Forough Malekmahmoudi -- Assessment of Engineering Properties of Clay Soils as Brick and Tiles Materials in Hamedan, West Of Iran / G.R. Khanlari, A. Namazi and Y. Abdi Lor -- Advances in Quantitative X-Ray Mineralogy: Mixed Crystals in Bauxite / Karsten Knorr and Reiner Neumann -- Quantitative XRD for Ore, Sinter, and Slag Characterization in the Steel Industry / Uwe König, Luciano Gobbo and Celeste Reiss -- Effect of Ground Glass Fineness on Physical and Mechanical Properties of Concrete / Aleksandrs Korjakins, Genadij Shakhmenko, Diana Bajare and Girts Bumanis.

Assessment of Quartzite Crystallinity Index by Ft-Ir / Michael V. Korovkin, Lyudmila G. Ananieva and Anastasia A. Antsiferova -- New Sample Preparation Developments to Minimize Mineral Segregation in Process Mineralogy / Rogerio Kwitko-Ribeiro -- Learnings From Five Years of On-Site Mla at Kennecott Utah Copper Corporation: (Myth Busters Through Quantitative Evidence ...) / Michael MacDonald, Ben Adair, Dee Bradshaw, Mick Dunn and Dewetia Latti -- Mineralogical Assessment of the Metamorphosed Broken Hill Sulfide Deposit, South Africa: Implications for Processing Complex Orebodies /

Craig R. McClung and Fanus Viljoen -- Deterioration of Concrete: Application of Stable Isotopes / Florian Mittermayr, Dietmar Klammer, Daniel Höllen, Stephan Köhler and Michael Böttcher, et al. -- Trace Elements of Pegmatitic Quartz and Their Regional Distribution in two Pegmatite Fields of Southern Norway / Axel Müller and Peter M. Ihlen -- Trace Element Content and Optical Cathodoluminescence of Kyanite / Axel Müller, Alfons M. van den Kerhof and Maarten A.T.M. Broekmans -- Weathering of Pyrrhotite to Szomolnokite / Alexander G. Nikiforov, Victor V. Onufrienok, Anatoly M. Sazonov and Alena V. Terehova -- Radioactive Elements in Bone Tissue From Freshwater Fish / Anton Nikolaev, Olga Frank-Kamenetskaya, Michael Zuykov and Elena Rosseeva -- Comparing the Mineralogical Characterisation of Iron Ore by Using Qemscan and Pta / Pejman Oghazi and Bertil I. Pålsson -- Influence of the Phase Composition of Pyrrhotites on Gold Content in Rocks / Victor V. Onufrienok, Anatoly M. Sazonov and Alena V. Terehova -- Application of Rapid X-Ray Diffraction (Xrd) and Cluster Analysis to Grade Control of Iron Ores / Mark Paine, Uwe König and Emma Staples -- Sulfide Oxidation and Mobilization of Arsenic in the Ketza River Mine Tailings / Dogan Paktunc, Yves Thibault and Chris Weisener -- Development of a Textural Acid Rock Drainage Index for Classifying Acid Formation / Anita Parbhakar-Fox, Mansour Edraki, Dee Bradshaw and Steve Walters -- Detection of the Deep Approach Fluid Channels in the Oil-and-Gas Basins Using the Geomagnetic Data (North and Norwegian Seas) / A. Petrova, I. Demina and M. Petrishchev -- Rare-Earth Element and Thorium Potential of Heavy Mineral Deposits Along the West Coast of South Africa With Special Reference to the Namakwa Sands Deposit / Carlo Philander and Abraham Rozendaal.

Influencing the Morphology of Gypsum / Christian Pritzel and Reinhard Trettin -- Co2 Sequestration During Interactions Between Fluid and Mafic to Intermediate Intrusive Rocks on Vannøya Island, West Troms Basement Complex, North Norway / Nadezhda S. Priyatkina, Kåre Kullerud, Steffen G. Bergh, Paul E.B. Armitage and Erling J.K. Ravna -- Synthesis and Hydration of High Reactive Brownmillerites $Ca_2(Fe_2\text{-XAlx})O_5$ / Bastian Raab and Herbert Pöllmann -- Mineralization of Iron Deposits from Dehbid Area, Fars Province, South Iran: Geochemical and Mineralogical Data / M.A. Rajabzadeh and S. Rasti -- Hyperspectral Imaging of Iron Ores / Erick R. Ramanaidou and Martin A. Wells -- Investigation on Microhardpan Texture using [mu]-CT and MLA / Dieter Rammlmair, Wiete Huebner and Maria Sitnikova -- Understanding Micro-Environment Development in Mine Tailings Using MLA and Image Analysis / Mostafa Redwan and Dieter Rammlmair -- Application of Edxrf Handheld to Optimize Exploration Strategies / Dieter Rammlmair, Mostafa Redwan and Samson A. Oke -- Rare Earth Element Mineralogy and its Recovery from the Neoproterozoic Riviera W-Mo Deposit, South Africa / Abraham Rozendaa and René Boshoff -- Process Mineralogy Delivering Significant Value at Anglo Platinum Concentrator Operations / Chris Rule and Robert P. Schouwstra -- Natural Carbon Nanostructuring Materials / Victor V. Ryabov, Victor A. Ponomarchuk, Anatoliy T. Titov and Dina V. Semenova -- Kinetic Studies of Hydration-Dehydration in Nano-Scale Interlayer Spaces for Smectite Clay Minerals / Kiminori Sato and Koichiro Fujimoto -- Olivine and Climate Change / R.D. Schuiling -- Staining of an Indoor Carrara Marble Floor at the Opera House Oslo Norway / Rune S. Selbekk, Elen Roaldset and Hans-Jørgen Berg -- Medieval Kirschsteinite-Bearing Iron Slags of the Develi-Yahyali Area (Kayseri), Turkey / W.E. Sharp and Steven K. Mittwede -- Högbomite Associated with Vanadiferous-Titaniferous Magnetite Bands at Bhaktarhalli, Nuggihalli Schist Belt, Western Dharwar Craton, Karnataka, India / P.V. Sunder Raju and R.K.W. Merkle -- Mineralogical Studies in Assisting Beneficiation of Rare Earth Element Minerals from Carbonatite Deposits / Wendy Thompson, Annegret Lombard, Eunice Santiago and Ashma Singh -- The Effect of Granulated Nickel Converter Matte Mineralogy on Phase-Specific Hardness and Associated Breakage Characteristics / Elton L. Thyse, Guven Akdogan, Jan H. Neethling, Ezra J. Olivier and William E. Goosen, et al.

Nanostructure of Biomineral Apatite of Modern Oceanic Phosphorites / Anatoly T. Titov, Gleb N. Baturin and Vladimir I. Zaikovskii -- Physicochemical Model of Calcium Phosphate Mineralization in Human Organism / Anatoly T. Titov and Piter M. Larionov -- Immobilisation of Uraniumvi in Cementitious Materials: Evidence for Structural Incorporation in Calcium-Silicate-Hydrates and Solid Solution Formation / J. Tits, X. Gaona, N. Macé, D. Kulik and T. Stumpf, et al. -- Characterization of the Network Structure of Alkali-Activated Aluminosilicate Binders by Single- and Double-Resonance 29si {27al} Mas Nmr Experiments / T.T. Tran, S.A. Bernal, D. Herfort and J. Skibsted -- Medieval Kirschsteinite-Bearing Iron Slags of the Develi-Yahyali Area (Kayseri), Turkey / W.E. Sharp and Steven K. Mittwede -- Microstructural Characterization of Fine Recycled Aggregates by Sem-Mla / Carina Ulsen, Henrique Kahn, Rafael Ribeiro de França, Daniel Uliana and Freud Seigfreid Campbell -- The Hötting Breccia: A Quaternary Building Stone: Its Formation, Quarries, Use and Material Technical Properties / Michael Unterwurzacher and Ulrich

Obojes -- Timan Bauxites: Distribution, Composition, Industrial Application (Komi Republic, Russia) / AV Vakhrushev and OB Kotova -- A New Cladding System Using Low Strength Sandstone in Precast Concrete Panels for Building Façades / Carmen Vielba-Cuerpo, Francisco Hernández-Olivares and Benito Lauret-Aguirregabiria -- Application of a Field Emission Mineral Liberation Analyser to the in Situ Study of Platinum-Group Element Mineralisation in the Merensky Reef of the Bushveld Complex, South Africa / Fanus Viljoen, Mike Knoper, Hariharan Rajesh, Derek Rose and Tiaan Greeff -- Spectroscopic Studies of the Influence of Multi-Walled Carbon Nanotubes on the Hydration of Tricalcium Silicate and Microstructure of Calcium Silicate Hydrate Phases / B. Weitzel, M.R. Hansen, T.L. Kowald, T. Müller and H.W. Spiess, et al. -- Petrological and Mineralogical Studies of the Sibin Bian-Stone, a Material for Making Acupuncture Tools in Ancient China / Xiande Xie, Fuya Wang, Zhenya Sun, Pui Leung Lo and Kit Chee Kong, et al. -- Mineralogy of Iron-Rich Sediments of Benavi Area in Kurdistan Region-Northern Iraq / Ali Taha Yassin and Mudhafar M. Mahmoud -- Sorption Capacity of Clay Minerals for Oil and Oil Products from Water Areas / Halyna Zadvernyuk -- Compare the Microstructure and Microenvironment of Dry and Wet Organic Montmorillonites / J.X. Zhu, Y.H. Ma, J.M. Wei, R.L. Zhu and P. Yuan, et al.

LC Subjects Mineralogy--Congresses.
Mines and mineral resources--Congresses.

Notes Includes bibliographical references and indexes.

Testing and evaluation of advanced building materials: selected, peer reviewed papers from the first national academic symposium on testing and evaluation of building materials (TEBM 2012), June 22-24, 2012, Shanghai, China

LCCN	2012540780
Type of material	Book
Main title	Testing and evaluation of advanced building materials: selected, peer reviewed papers from the first national academic symposium on testing and evaluation of building materials (TEBM 2012), June 22-24, 2012, Shanghai, China / edited by Wu Yao.
Published/Produced	Durnten-Zurich, Switzerland: Trans Tech Publications Ltd, [2013] ©2013
Description	v, 282 pages: illustrations (black and white); 24 cm.
Rights advisory	Current copyright fee: GBP18.00 50\0.
ISBN	9783037855775 (paperback) 3037855770 (paperback)
LC classification	TA410 .T39 2013
Related names	Yao, Wu, editor.
Contents	Machine generated contents note: 29Si NMR Characterization of Silica Tetrahedron in the Silica Fume Simulate Hydration / C. Qian -- Investigation of Early Cement Paste with 1H Low-Field NMR / P.Q. Yang -- The Application of Freezing-Melting Hysteresis in Hardened White Cement Paste / L. Zhou -- An Experimental Study of Water in Pore System of Hardened Cement Paste by Magnetic Resonance / W.C. Yuan -- Quantitative Characterization of Hydration of Cement Pastes by Rietveld Phase Analysis and Thermoanalysis / W. Yao -- Kinetic Study of Portland Cement Hydration with Ground Penetrating Radar / W.R. Zhang -- Study on the Hydration Kinetics of Portland Cement / Y.Q. Wei -- Relationship between Internal Relative Humidity and Autogenous Shrinkage of Cement Paste with Supplementary Cementitious Materials (SCM) / Q.Q. Yan -- A Testing Device for

Humidity-Control Performance of Pervious Concrete / X.F. Lv -- An Evaluation of Shrinkage Model Based upon Microstructure of Blended Cement Pastes / D. Jin -- Early-Age Free Shrinkage of Mortars with Different Dosages of EVA Redispersible Powder / P.M. Wang -- Shrinkage of Blended Cement Pastes with Mineral Additions / D. Jin -- Relative Humidity of Blended Cement Pastes in Sealed during Hydration / Y. Chen -- Effect of Hollow Glass Microsphere on Performance of Foam Concrete / X.F. Yan -- Effects of Rubber Powder and Fly Ash on Mechanical Properties of Recycled Mortars / K. Gao -- Micromechanical Properties of Calcium Silicate Hydrate / L. He -- Probing Nanostructure of Calcium Silicate Hydrate by AFM and Nanoindentation / Q. Gao -- Study on the Unhydrated Cement Grain/C-S-H Gel Interface in Cement Paste by Use of Nano-Scratch Technique / J. Xu -- Temperature Sensitive Properties of Hybrid Carbon Nanotube/Carbon Fiber Cement-Based Materials / J.Q. Zuo -- Research on Optimizing the Electrical and Mechanical Properties of Carbon Fiber Reinforced Cement / W. Yao -- Mechanical Property of Hybrid Steel Fiber Reinforced Cement-Based Composites / S. Dong -- Enhancing the Thermoelectric Properties in Carbon Fiber/Cement Composites by Using Steel Slag / J.J. Qin -- Seismic Behaviour of RC Columns Strengthened with Steel Bar/Wire Mesh Mortar / G.J. Xiong -- Carbonation Profile of Cement Paste and Concrete Established with Micro-Hardness Analysis / S.Z. Zhang -- Effect of Activated Water Treatment Sludge on Carbonation of Mortar / Y. Chen -- Effect of Fly Ash on Resistance to Sulfate Attack of Cement-Based Materials / L.W. Mo -- Study of Durability Analysis and Evaluation Model for Existing Concrete Structure in Coastal Areas / H. Zhang -- The Effect of Fly Ash on TSA of Cementitous Material: Based on Three Years

Results / J. Zhang -- Study of Mechanical Force on Coal Gangue Reactivity / L. Li -- The Effect of Electrochemical Chloride Extraction Combining Ultrasonic on Steel-Reinforced Mortars / W. Yao -- Tentative Study on Sonoelectrochemical Chloride Extraction from Mortar / X.M. Xing -- Analysis of Bond and Anchorage Performances of Helical and Twisted Reinforcement Material / C.L. Wang -- Research on Modification of Steady State Migration Test for Cementitious Materials / L.P. Wu -- Study on Hydration Degree of Portland Cement-Slag Complex Binders / P.Y. Yan -- Study on Pore Structure Characterization of Concrete at Different Ages by Thermoporometry / N. Zhang -- Determining the Contact Angle of Hardened Cement Paste Using Thin Layer Wicking Method / J.P. Liu -- Solubility Behavior of the Hydration Products in the Pore Solution of Hydrated Cement Pastes / M.J. Wu -- Effect of Metakaolin on the Physical Properties and Setting Time of High Performance Concrete / Y. Han -- The Effect of Curing System on Mechanical Properties of Desulphurization Gypsum-Slag Composite Binder / H.G. Qin -- The Influence of Curing Temperature on the Coordination of the Expansion and Strength of High Strength Expansive Concrete / X.Q. Wang -- Preparation of MgO- and CaO-Bearing Expansive Agent Used for Cement-Based Materials / M. Deng -- Study on Compatibility of Modified the Cement-Based Composite Biomass Materials by Agent / X.M. Wang -- The Brief Analysis of Photovoltaic Insulating Glass in the Process Service / C.G. Wei -- Phosphate Bonding: A New Method for Using Large Volume of Fly Ash / H. Lu -- The Preparation and Pozzolanic Activity of Metakaolin Admixtures / M. Li -- Influence of Steel Slag on the Workability of Concrete / B. Zhang -- Study on the Influence Factors of the Dispersion of Carbon Nanotubes in Aqueous Solution / M.K. Zhou --

	Properties of Cement Mortars Mixed with SiO2 and CaCO3 Nanoparticles / Y.F. Meng -- Alkali Activity of Granite Aggregates and Control in the Concrete of the Three Gorges Project / L. Xiao -- Preparation of AgBr Nanowire Arrays in Porous Anodic Aluminium Oxide Template by Paired Cell Deposition / J.W. Hou -- Influence and Mechanisms of In Situ Toughening Telechelic Polymer on the Microstructure of Concrete / J.X. Hong -- Research on the Application of Liquid Permeating Method to Silica Sol Stability / W.H. Huang -- Environmentally Friendly Cellular Concrete for Wall Insulation / Y. Chen.
LC Subjects	Building materials--Congresses. Building materials--Testing--Congresses.
Other Subjects	Building materials. Building materials--Testing.
Subject keywords	Advanced building materials; TEBM;
Form/Genre	Conference proceedings.
Notes	Includes bibliographical references and index.
Series	Key engineering materials, 1013-9826; volume 539

Water treatment processes

LCCN	2011033645
Type of material	Book
Main title	Water treatment processes / editor, Kostas Demadis.
Published/Created	Hauppauge, N.Y.: Nova Science Publishers, c2012.
Description	xvii, 487 p.: ill.; 26 cm.
ISBN	9781621003526 (hbk.) ebook
LC classification	TD430 .W377 2012
Related names	Demadis, Kostas.
Contents	An overview of electrochemical treatment processes in environmental engineering: definitions, electrolytes and electrodes -- An overview of electrochemical treatment processes in environmental engineering: previous studies on electrochemical treatment processes (1973- 2004) -

- An overview of electrochemical treatment processes in environmental engineering: modern studies on electrochemical treatment processes (2005- 2010) -- Corrosion and fouling aspects -- Application of alkylamines as a base for anticorrosive and antiscaling formulations -- Gas chromatographic separation and mass spectrometric identification of corrosion inhibiting long-chain primary alkyl amines and alkyl diamines applied in water treatment in the power industry -- Microbiologically influenced corrosion -- Gypsum scaling in a flow system -- Inhibition of gypsum scale formation -- Removal techniques -- New advances in natural coagulants for water purification: moringa oleifera seed extract -- Hydrothermal synthesis of diatomaceous earth covered with calcium silicate hydrate gels for dye removal applications -- Multi-component adsorption equilibria of PB2+ and CD2+ onto powdered corn cobs (PCC) -- Treatment of heavy metals from industrial wastewaters using hollow fiber supported liquid membrane -- Organic/inorganic hybrid materials based on functional dendrimers and hyperbranched polymers for water purification -- Electrochemical techniques -- Heterogeneous fenton-like processes for the treatment of industrial wastewaters: a review with special attention to iron-containing silica catalysts Botas -- An overview of electrochemical: technique for decolouring aqueous solutions with optimization and economics -- A statistical evaluation of selected factors that influence efficacy of fenton process in wastewater treatment -- Sulphate removal from aqueous solutions by using electrochemical techniques -- Public health -- Technologies for early warning and detection of cyanobacterial harmful algal blooms leading to integrated source water protection -- Statistical treatments of non-linear regression method with BOD kinetics in wastewater treatment techniques.

LC Subjects Water--Purification.

Notes Includes bibliographical references and index.

Index

D

E

F

G

H

I

K

L

S

T

W

X